HoloLens Blueprints

Experience the virtual and real worlds coming together with HoloLens

Abhijit Jana
Manish Sharma
Mallikarjuna Rao

BIRMINGHAM - MUMBAI

HoloLens Blueprints

First published: June 2017

Production reference: 1160617

Published by Packt Publishing Ltd.
Livery Place
35 Livery Street
Birmingham
B3 2PB, UK.
ISBN 978-1-78728-194-3

www.packtpub.com

Credits

Authors
Abhijit Jana
Manish Sharma
Mallikarjuna Rao

Copy Editor
 Dhanya Baburaj

Reviewer
Michael Washington

Project Coordinator
Ritika Manoj

Commissioning Editor
Smeet Thakkar

Proofreader
Safis Editing

Acquisition Editor
Denim Pinto

Indexer
Tejal Daruwale Soni

Content Development Editor
Mohammed Yusuf Imaratwale

Graphics
Jason Monterio

Technical Editor
Akansha Bathija

Production Coordinator
Arvindkumar Gupta

About the Authors

Abhijit Jana works with Microsoft as a development consultant as part of Microsoft Services. As a consultant, his job is to help customers design, develop, and deploy enterprise-level secure solutions using Microsoft technologies. Apart from being a former Microsoft MVP (Most Valuable Professional), he is a speaker and author as well as an avid technology evangelist. He has delivered sessions at prestigious Microsoft events, such as TechED, Web Camps, Azure Camps, Community TechDays, Virtual TechDays, DevDays, and developer conferences. He loves to work with different .NET communities and help them with different opportunities.

He is a well-known author and has published many articles on various .NET community sites. You can follow him on Twitter at @abhijitjana. He has authored the book *Kinect for Windows SDK Programming Guide* (ISBN: 1849692386 ISBN 13: 9781849692380).

Abhijit lives in Hyderabad, India, with his wife, Ananya, and a beautiful little angel, Nilova.

Manish Sharma works with Microsoft as a solution architect as part of Microsoft Services. As, a solution architect, he is responsible for leading large enterprise-transformational engagements defining technology/solution roadmaps, conducting architecture and technical evaluations of enterprise engagements, and architecting multimillion-dollar solution developments, and maintenance of outsourcing management. He is also a technology evangelist and speaker at prestigious events, such as Microsoft TechEd on the latest and cutting edge technologies, such as HoloLens, Internet of Things, connected cars, and cloud technologies such as Azure.

Mallikarjuna Rao works with Microsoft as UX Designer/3D mixed reality artist as part of Microsoft Services. He has 15 years of experience in the animation industry, with around 12 years' international exposure.

He has worked as a senior 3D asset supervisor for feature films, television series, commercials, and game projects. Mallikarjuna has worked on pivotal projects for big names in the industry, such as Walt Disney Pictures, Disney Junior, Warner Bros Pictures, Yash Raj Films, and UTV Motion Pictures, which put him under the limelight at the international level. He has an in-depth knowledge of the 3D Asset production pipeline and UX design principles.

Acknowledgments

Writing this book would not have been possible without help of many people. We had a wonderful time while writing, which was mainly due to the skills support, dedication and motivation of people around us.

First of all, we are extremely thankful to Kundan Prakash, Ankur Agarwal and Shanmugadas Chembakassery Sivadasan, Suresh Viswanath for their continuous support and motivation from the time we started writing this book.

We are deeply thankful to the entire team at Packt Publishing, including Tushar, Parul, Narendra, Ritika, Smeet, Denim, Safis, Dhanya, Tejal, Jason and Arvind. We would like to extend our thanks to Content Development Editor, Yusuf and our Technical Editor Akansha. They have been very supportive and helpful from the beginning. Thank you all for your effort and dedication.

A sincere thanks to Michael Washington for their insightful and deep technical review. He helped us to identify and fill the gaps and improve the quality of the book.

We would like to thank Charul Pant and Akshay Dixit for their offline review of the book and sharing their feedback. It was extremely helpful to get some feedback from avid reader and technology geek like you people. Thank you very much for your time and help us improving the book over the period.

We spent time in writing when we should have sleeping, spending time with family or playing with our kids. We wouldn't have never been able to write this book without support from our family. Be it late night working or meeting over weekends, they supported us every bit and pieces.

About the Reviewer

Michael Washington is an ASP.NET, C#, and Visual Basic programmer. He has extensive knowledge of process improvement, billing systems, and student information systems. He is a Microsoft MVP. He has a son, Zachary, and resides in Los Angeles with his wife, Valerie.

Disclaimer

The opinions in this book are purely our personal opinions and do not reflect in any way the opinion of our employers.

Customer Feedback

Thanks for purchasing this Packt book. At Packt, quality is at the heart of our editorial process. To help us improve, please leave us an honest review on this book's Amazon page at http://www.amazon.in/dp/1787281949.

If you'd like to join our team of regular reviewers, you can e-mail us at customerreviews@packtpub.com. We award our regular reviewers with free eBooks and videos in exchange for their valuable feedback. Help us be relentless in improving our products!

www.PacktPub.com

For support files and downloads related to your book, please visit www.PacktPub.com.

Did you know that Packt offers eBook versions of every book published, with PDF and ePub files available? You can upgrade to the eBook version at www.PacktPub.com and as a print book customer, you are entitled to a discount on the eBook copy. Get in touch with us at service@packtpub.com for more details.

At www.PacktPub.com, you can also read a collection of free technical articles, sign up for a range of free newsletters and receive exclusive discounts and offers on Packt books and eBooks.

https://www.packtpub.com/mapt

Get the most in-demand software skills with Mapt. Mapt gives you full access to all Packt books and video courses, as well as industry-leading tools to help you plan your personal development and advance your career.

Why subscribe?

- Fully searchable across every book published by Packt
- Copy and paste, print, and bookmark content
- On demand and accessible via a web browser

"I dedicate this book to my parents, my lovely wife Ananya and my little angle Nilova" -
Abhijit Jana

"I dedicated this book to my parents, my wife Joohi and lovely son Atharv" - *Manish Sharma*

*"I would like to dedicate this book to my parents, my wife Sunitha and kids Sai Thanish, Rudhvi
sriman"* -*Mallikarjuna Rao*

Table of Contents

Preface

Digital reality brings immersive experiences, such as transporting you to different world or places, making you interact within those immersive, mixing digital experiences with reality, and ultimately, opening new horizons to make you more productive.

The spectrum and scenarios of digital reality are huge; to understand them better, they are broken down into three different categories: Virtual Reality (VR), Augmented Reality (AR), and Mixed Reality (MR).

Microsoft HoloLens is the most powerful mixed reality personal computing Windows device. It allows you to place your digital content in the real physical world, where it matters most to you. This is what is called an MR device, one that tries to blend the real and digital worlds seamlessly and provides new ways to visualize the world beyond the screens.

The boundaries between the real and the digital world start to blur, and you can interact with the digital world in a very similar way to how you would interact with the real, physical world. What is it about HoloLens that makes it stand out from the other devices in this genre? It is the fact that Microsoft HoloLens is the first fully untethered, self-contained computer running on Windows 10. Yes, that's right: no hassle of wires, additional cables, or connections to a computer or additional devices. It is hands-free, self-contained device with a built-in battery.

This book is mainly focused on the building mixed reality application using Microsoft HoloLens. This book doesn't require any prior knowledge about the holographic platform from the reader, and its strength is the simplicity in which the concepts have been presented using sketching, architecting, code snippets, a step-by-step process, and detailed descriptions. This book covers the following:

- Understanding what mixed reality is and the possibilities of it
- Understanding the process of developing enterprise holographic applications using HoloLens
- Learning the application and interaction model for holographic apps
- Understanding holographic application integration possibilities within Line of Business application using Azure
- Learning by realistic integrated project examples
- Learning end-to-end complex integrated solutions using holographic app and Azure

What this book covers

Chapter 1, *Digital Reality - Under the Hood*, begins your mixed-reality journey by explaining different types of digital realities, their differences, and usage scenarios. Also, it will explain the different market players and how they are different from each other based on their products. This chapter will also introduce you to the immersive HoloLens device and show you how to take a step further with the device.

Chapter 2, *HoloLens – the Most Natural Way to Interact*, introduces HoloLens as a hardware device. You will get an insight into the different components that make up HoloLens and the technology behind the device, which makes the components work together. This chapter also gives an overview of holograms and ways to interact with holograms with different interaction models. You will also become familiar with HoloLens Application Model and take a step toward developing holographic applications by understanding the team structures and roles required.

Chapter 3, *Explore HoloLens as Hologram - Scenario Identification and Sketching*, dives into the development process of developing your first holographic application. You will learn about identifying the scenario and the interaction models required within that scenario, such as how the user will interact with gaze, gesture, and other interaction models. You will also learn how to identify 3D Assets and take a quick overview of creating new 3D Assets.

Chapter 4, *Explore HoloLens as Hologram - Developing Application and Deploying on Device*, further enhances the scene developed in the preceding chapter and adds new life to it by developing different interaction models using scripting, building interaction model scripts, and testing them in Unity3D and holographic emulators. Finally, you deploy the application on a device and test your first holographic application, and visualize hologram interactions with the real world.

Chapter 5, *Remote Monitoring of Smart Building(s) Using HoloLens Scenario Identification and Sketching*, teaches you how to identify and design real-world integrated scenarios. You will integrate Azure IoT solutions with holographic application and visualize the data feed coming in near real time from the cloud through holograms. You will learn about integration scenario options, architecture, and the interaction models required within this scenario.

Chapter 6, *Remote Monitoring of Smart Building(s) Using HoloLens - Developing Application and Deploying on Device,* takes you through integration scenario identified in the preceding chapter, shows you how to implement and develop the holographic application, and explain how to integrate it with IoT system hosted on Azure. You will learn how to pull the data from the IoT system and render 3D assets on the holographic application based on the data. In this chapter, you will build interaction model scripts, test them in Unity and emulators, and finally deploy them on a device and test your first integrated application.

Chapter 7, *Build End-to-End Retail Solution - Scenario Identification and Sketching,* will teach you how to identify and design end-to-end retail solutions. You will sketch a solution to design your home appliances with virtual holograms. Then, you will place an order directly from the holographic application.

Chapter 8, *Build End-to-End Retail Scenario - Developing Application and Deploying on Device,* will take the retail scenario identified in the preceding chapter and implement and develop its holographic application. It will integrate it with backend services hosted on Azure. You will learn how to pull the data from the backend system, render 3D Assets on the fly, and place an order within holographic application. Here, you will build interaction model scripts, test them in Unity and emulators, and finally deploy them on a device and test your end-to-end retail application.

Chapter 9, *Possibilities,* outlines the new possibilities and scenarios with mixed reality and HoloLens.

Chapter 10, *Microsoft HoloLens in Enterprise,* looks at managing enterprise devices, such as adding them onto enterprise domains, managing updates, application distribution, and other device management features, such as managing devices through Microsoft Intune.

What you need for this book

The basic requirements for this book are as follows:

- Unity3D - 5.5 or higher
- Visual Studio 2015 (Update 3) or higher
- HoloLens emulator
- Optional requirements:
    - Autodesk Maya
    - Adobe Photoshop
    - HoloLens device

Refer to Chapter 03, *Explore HoloLens as Hologram - Scenario Identification and Sketching*, for detailed information on the development environment setup.

Who this book is for

This book is intended to help experienced developers and designers start working on mixed reality development using Microsoft HoloLens. This book is also useful for entrepreneurs who are planning to build a team of the following:

- Developers/professionals
- Designers
- Entrepreneurs who are planning to build a team for mixed reality development

This book uses C# and Unity3D Scripting (C#) in the examples, so you need to know the basics of C# and scripting within Unity3D. You should be familiar with Visual Studio IDE and Unity Editor as well. You don't have to know anything about the HoloLens as a device.

Conventions

In this book, you will find a number of text styles that distinguish between different kinds of information. Here are some examples of these styles and an explanation of their meaning.

Code words in text, database table names, folder names, filenames, file extensions, pathnames, dummy URLs, user input, and Twitter handles are shown as follows: "By default, the FloorInteractionManager class is inherited from MonoBehaviour. MonoBehaviour is the base class from which every Unity script derives. In the next step, implement IFocusable interface in the FloorInteractionManager class. "

A block of code is set as follows:

```
//On Start, create WWW object and link it with Steaming URL
void Start() {
    webObject = new WWW(videoUrl);
    guiTexture = GetComponent<GUITexture>();
    guiTexture.texture = webObject.movie;
}
```

New terms and **important words** are shown in bold. Words that you see on the screen, for example, in menus or dialog boxes, appear in the text like this: "Once the basic design is done, tag all the text controls as **BuildingTitle**, **FireText**, **Temp Text**, and **SmokeText** for **Main Title**, **Fire**, **Temperature**, and **Smoke** elements, respectively."

Warnings or important notes appear in a box like this.

Tips and tricks appear like this.

Reader feedback

Feedback from our readers is always welcome. Let us know what you think about this book-what you liked or disliked. Reader feedback is important for us as it helps us develop titles that you will really get the most out of.

To send us general feedback, simply e-mail `feedback@packtpub.com`, and mention the book's title in the subject of your message.

If there is a topic that you have expertise in and you are interested in either writing or contributing to a book, see our author guide at `www.packtpub.com/authors`.

Customer support

Now that you are the proud owner of a Packt book, we have a number of things to help you to get the most from your purchase.

Downloading the example code

You can download the example code files for this book from your account at `http://www.packtpub.com`. If you purchased this book elsewhere, you can visit `http://www.packtpub.com/support` and register to have the files e-mailed directly to you.

You can download the code files by following these steps:

1. Log in or register to our website using your e-mail address and password.
2. Hover the mouse pointer on the **SUPPORT** tab at the top.
3. Click on **Code Downloads & Errata**.
4. Enter the name of the book in the **Search** box.
5. Select the book for which you're looking to download the code files.
6. Choose from the drop-down menu where you purchased this book from.
7. Click on **Code Download**.

Once the file is downloaded, please make sure that you unzip or extract the folder using the latest version of:

- WinRAR / 7-Zip for Windows
- Zipeg / iZip / UnRarX for Mac
- 7-Zip / PeaZip for Linux

The code bundle for the book is also hosted on GitHub at `https://github.com/PacktPubl ishing/HoloLens-Blueprints`.We also have other code bundles from our rich catalog of books and videos available at `https://github.com/PacktPublishing/`. Check them out!

Downloading the color images of this book

We also provide you with a PDF file that has color images of the screenshots/diagrams used in this book. The color images will help you better understand the changes in the output. You can download this file from `https://www.packtpub.com/sites/default/files/downloads/HoloLensBlueprints_Color Images.pdf`.

Errata

Although we have taken every care to ensure the accuracy of our content, mistakes do happen. If you find a mistake in one of our books-maybe a mistake in the text or the code- we would be grateful if you could report this to us. By doing so, you can save other readers from frustration and help us improve subsequent versions of this book. If you find any errata, please report them by visiting `http://www.packtpub.com/submit-errata`, selecting your book, clicking on the **Errata Submission Form** link, and entering the details of your errata. Once your errata are verified, your submission will be accepted and the errata will be uploaded to our website or added to any list of existing errata under the Errata section of that title.

To view the previously submitted errata, go to `https://www.packtpub.com/books/content/support` and enter the name of the book in the search field. The required information will appear under the **Errata** section.

Piracy

Piracy of copyrighted material on the Internet is an ongoing problem across all media. At Packt, we take the protection of our copyright and licenses very seriously. If you come across any illegal copies of our works in any form on the Internet, please provide us with the location address or website name immediately so that we can pursue a remedy.

Please contact us at `copyright@packtpub.com` with a link to the suspected pirated material.

We appreciate your help in protecting our authors and our ability to bring you valuable content.

Questions

If you have a problem with any aspect of this book, you can contact us at `questions@packtpub.com`, and we will do our best to address the problem.

1
Digital Reality - Under the Hood

Welcome to the world of Digital Reality. The purpose of Digital Reality is to bring immersive experiences, such as taking or transporting you to different world or places, make you interact within those immersive, mix digital experiences with reality, and ultimately open new horizons to make you more productive. Applications of Digital Reality are advancing day by day; some of them are in the field of gaming, education, defense, tourism, aerospace, corporate productivity, enterprise applications, and so on.

The spectrum and scenarios of Digital Reality are huge. In order to understand them better, they are broken down into three different categories:

- **Virtual Reality** (**VR**): It is where you are disconnected from the real world and experience the virtual world. Devices available on the market for VR are Oculus Rift, Google VR, and so on. VR is the common abbreviation of Virtual Reality.
- **Augmented Reality** (**AR**): It is where digital data is overlaid over the real world. Pokémon GO, one of the very famous games, is an example of this globally. A device available on the market, which falls under this category, is Google Glass. Augmented Reality is abbreviated to AR.
- **Mixed Reality** (**MR**): It spreads across the boundary of the real environment and VR. Using MR, you can have a seamless and immersive integration of the virtual and the real world. Mixed Reality is abbreviated to MR.

This book is mainly focused on developing MR applications using Microsoft HoloLens devices.

Although these technologies look similar in the way they are used, and sometimes the difference is confusing to understand, there is a very clear boundary that distinguishes these technologies from each other. As you can see in the following diagram, there is a very clear distinction between AR and VR. However, MR has a spectrum, which overlaps across all three boundaries of real world, AR, and MR.

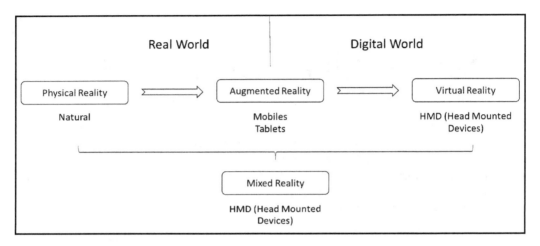

Digital Reality Spectrum

The following table describes the differences between the three. Later in the chapter, we will go through the details of each one of them.

Virtual Reality	Augmented Reality	Mixed Reality
• Complete Virtual World • User is completely isolated from the real world • Device examples: Oculus Rift and Google VR	• Overlays Data over the real world • Often used for mobile devices • Device example: Google Glass • Application example: Pokémon GO	• Seamless integration of the real and virtual world • Virtual world interacts with real world • Natural interactions • Device examples: HoloLens and Meta

Let's explore VR, AR, and MR in more detail.

Virtual Reality – what it is?

The dictionary definition of VR reads, "*a realistic and immersive simulation of a three-dimensional environment, created using interactive software and hardware, and experienced or controlled by movement of the body*". However, we will say that there is one and only one primary purpose of the VR, that is, to make you believe that you are somewhere else, for example, in a gaming environment, a war zone, or touring some city.

Virtual Reality illustration

Realizing Virtual Reality

A person wearing a VR device/headset will be able to view the virtual environment, move around wearing it, and interact with items within that virtual environment. Most Virtual Realities are created using a head-mounted helmet or a set of goggles, which are given the generic name: **head-mounted display (HMD)**. HMD is usually connected to a computer or a smartphone, which does all the 3D rendering work, and HMD is just used to display that rendered 3D content. HMD completely covers user's view, by covering both eyes. As a result, the user is completely cut off from the outside world and completely focused on the virtual or digital world.

HMDs basically consists of stereoscopic displays and motion tracking hardware. The way HMDs implement stereoscopic displays is by generating different image for each eye, which results in generating the illusion of depth. Within HMDs, motion tracking hardware mostly consists of a gyroscope and accelerometer to measure motion/position changes. This helps in simulating real-world experiences.

Virtual Reality in the field

There are various implementations of VR in diverse fields currently, and numerous new ones are coming up every day. It will be a very interesting domain for the next few years, where we will globally see lots of new fields picking up VR and merging it with other upcoming technologies. One such technology, which we can think of, is **Artificial Intelligence (AI)**; Are we talking about the movie "Matrix" here?

While it's too early to predict possibilities in the future, let's explore some of the current fields where VR is being implemented. They are as follows:

- **Gaming**: Gaming was one of the first fields to pick up, start using, and commercialize VR. Apart from single-user VR games, multi-user online VR games are also available today. With the availability of low-cost devices, such as Google VR and smartphones, VR games/applications are within the reach of everyone.
- **Tourism**: Tourism is another field that has picked up VR. VR is used for generating stereoscopic 360-degree panoramic views, where the user interact with the virtual place and explores it. So, no more traveling or air-travel; just wear your VR device and explore new places every day.
- **Education**: Distant learning, visualization, and interactive teaching a few global examples of VR implementations within the education field. Earlier, teachers used to describe the scene and students used to imagine the possibilities. Now, they just visualize new possibilities and interact with them, and spend more time in thinking what's next.
- **Architecture**: Architecture is a field where VR has made life easier for the architects by making it easy to create prototypes. Now, they do not have to create cardboard prototypes or sketches to convey their thoughts and ideas. VR has made it possible for them to share ideas early in the cycle of development and receive early feedback. This saves lot of time in rework.

- **Enterprise productivity:** Especially, for enterprise training scenarios, VR is used very frequently. Scenarios in consideration can be simulating war situations for the armed forces, flight simulation for pilots, medical diagnostics simulations for doctors, and many others.
- **Web content:** With the support of WebGL technology by most of the latest browsers available in the market, rendering Virtual 3D content has become very easy. A lot of commercial websites for automobile companies, for example, have started publishing their automobile features in the form of 3D content, which has made it very interactive for end users to visualize the content.

Augmented Reality – what it is?

AR is all about bringing digital information and overlaying it over the real environment. The only difference from VR is that it creates a totally artificial environment around you, whereas AR uses the environment around you and overlays the digital information over it. For VR, you will require a VR device, but for AR it can be achieved by simply using a smartphone, tablet, or dedicated VR devices.

Visualization techniques for Augmented Reality

AR can be categorized into three different types based on the display types:

- Screen-based
- HMD-based
- Projection-based

Screen-based

A common example of screen-based AR is overlaying digital information over smartphones or tablet camera displays. For example, you switch on and point your smartphone/tablet camera over an object, and the application recognizes that object and overlays that object information, such as the price or description, as digital information over the object image.

Another example of screen-based AR is the video game Pokémon GO, in which, based on the user's location and direction, digital characters are overlaid over video images.

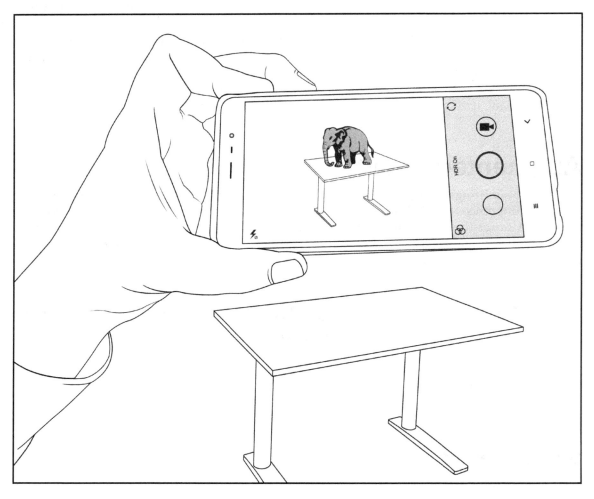

User is viewing augmented object using smartphone

In the preceding figure, a user is viewing an augmented object, that is, the elephant's digital object is overlaid over the video frame. This is an example of screen-based AR.

Augmented Reality head-mounted globally displays

Using AR HMD devices, digital information is overlaid directly over the user's view of the real world. So, there is no need to hold any screen to view the digital information. Digital information is directly rendered over the view area of the user's eyes.

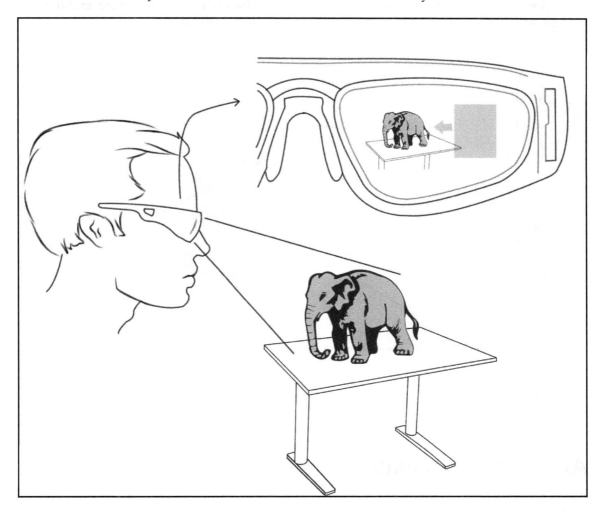

HMD with information overlaid over the real view

In the preceding image, the user is using a head mounted AR device to view the physical element in front of him. Within the view of the user, the AR device embeds information about the physical object and provides them with a more immersive experience.

Projection-based

User projection-based AR, a projection is rendered on the target surface itself. This target surface could be anything, such as building, person, room, and so on. To render this kind of projection, the system needs to know the exact dimensions of the target surface, and then using single/multiple projectors, it renders the projection on the target:

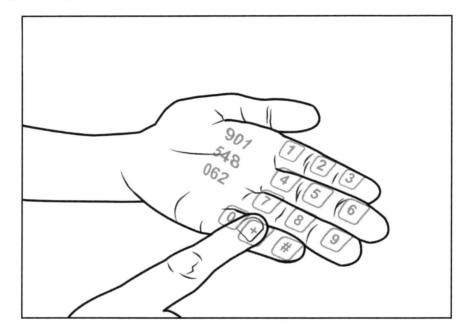

Phone dialer pad projection over the hand

In the preceding image, the user is viewing a projection of the phone dialer on the palm of their hand.

Augmented Reality in the field

Applying AR is quite different from applying VR. AR applications are more focused on integration scenarios with the real world. Some are as follows:

- **Gaming**: With the release of the Pokémon GO game by Niantic, the demand for AR games has picked up drastically in the market. A lot of companies are coming up with AR games and launching them.

- **Architecture/construction/archaeology**: AR can be used to visualize completed buildings for in-progress or upcoming building construction. Digital architecture/building images can be overlaid over the real view of the property or ground.

- **E-Commerce**: Businesses can't reach each customer with demo-able physical product, and opening showrooms is every city is a costly business, especially for start-ups and newcomers in the market. So, reaching out to customers through AR, online furniture retail companies, for example, allows users to consume AR through smartphones, let them visualize their furniture products, and enable them to design their house interiors.

- **Education**: Traditional education systems require hands-on instructions and real/prototype equipment to explain it better to students. With AR, teachers and students can visualize the same equipment in virtual mode, with very similar training instructions as with an actual device. Another implementation of AR in the education system is distance learning, where students and a teacher very far away, can use VR devices for interactions and virtual classrooms.

- **Medical**: In the medical field, there are different imaging techniques for various requirements, such as X-ray, ultrasound, and **magnetic resonance imaging (MRI)**, but there is no consolidated view for medical practitioners. AR could be used for these scenarios, where the output from different imaging techniques could be overlaid over the patient and give a consolidated view to medical practitioners.

- **Industrial design**: AR is being used for designing, sharing ideas, and brainstorming design views among different designers and architects and supplies quick feedback and brainstorming cycles. Earlier, the same process used to take a long time, as designers used to create physical prototypes and then discuss them.

- **Travel/navigation/tourism**: AR is also used for developing navigational applications; for example, travel-related digital information is overlaid on the vehicle windscreen, which helps the driver in real-time navigation without looking at any other device, such as a GPS or smartphone. AR is also used to develop travel-related applications, which help the user with location-specific information of the place where the user is currently placed, such as historical information about tourist places, or information about nearby restaurants and cuisines.

Mixed Reality – what it is?

We talked about VR and AR. Now, think of MR as a hybrid of VR and AR. Using MR, a user can view the real world just like in AR, but can also view virtual objects as in VR. The difference with MR is that those virtual objects are referenced or anchored with the real space. That means that the virtual object is interacting within the real world.

For example, suppose you have a physically real table in front of you, and now using MR you are viewing a virtual ball; if you drop a virtual ball object on a table, in MR, that virtual ball will interact with the physical table. This means that it will bounce on the physical table, and roll off from the table to the ground.

The immersive experience of the virtual world with the real world is so strong in MR that it is very hard to break that illusion.

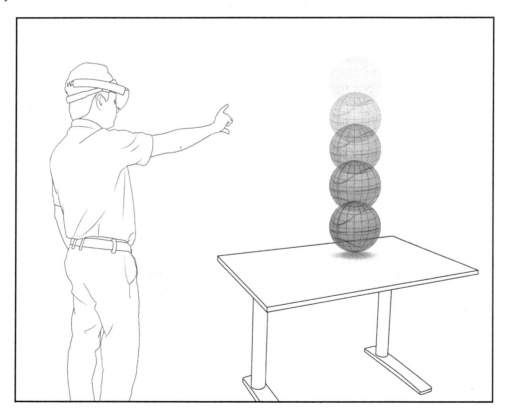

A user viewing a MR object in real space

At the time of writing this book, there are very few MR devices available on the market; Microsoft HoloLens is one of them. This book is focused on developing the MR experience, using HoloLens.

Mixed Reality in the field

MR is still evolving, and lot of different industries are still exploring possibilities with it. The following are some of them, which represent early progress:

- **Education and training**: MR has already started transforming the way education and industrial training are delivered. Companies have started using MR as a medium for delivering readiness/training content to participants.
- **Exploring new markets**: MR is giving an opportunity to enterprises to rethink their marketing strategy and target new customers. Few popular automobile companies have started using MR device in their specialized car showrooms. With this, they can target consumers early in the sales cycle, even before the first model of the car is available.
- **Healthcare**: The possibilities for MR in the healthcare industry are enormous. Physicians can use MR devices to visualize consolidated reports, and these devices can also be used as a guidance device during operations or diagnosis.
- **Gaming:** The gaming industry may become one of the biggest consumers of MR devices. There are so many possibilities for MR games that can interact with the real world.

Digital Reality and the current market place

We learned about VR, AR, and MR; let's see what devices are currently available on the market to explore all these different experiences:

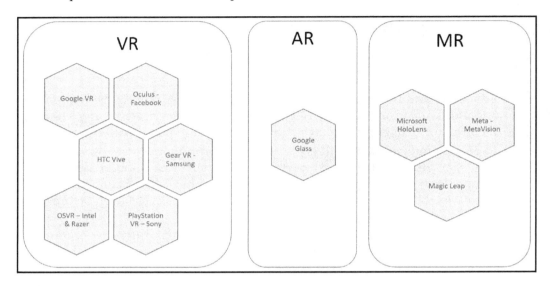

Digital Reality device family

- **Oculus**: Owned by Facebook, the first in the industry to launch a VR device
- **HTC Vive**: A VR device launched by the Taiwanese phone manufacturer HTC
- **Sony PlayStation VR**: A VR device that works with Sony PlayStation, but also works with any personal computer
- **Google VR**: This is a Google VR device, and is used along with Android smartphones
- **Gear VR**: This is a Samsung VR device, and is used along with Galaxy smartphones
- **OSVR**: An open source VR device, started with the support of companies such as Intel and Razer
- **Google Glass**: An AR device launched by Google
- **HoloLens**: Microsoft MR wireless and wearable device
- **Met**: Mixed or Augmented Reality device, which projects virtual images in the wearer's field of vision
- **Magic Leap**: MR device, still under development

Summary

We have seen in this chapter the different types of reality solutions available on the market and their possible usage. We have learned about how these digital realities are similar to each other and how they differ. We have also explored several field applications of these technologies. We have seen the different sets of devices available in the current market place. The knowledge gained from this chapter will help you grasp the subject discussed in subsequent chapters.

Further on within this book, we will focus on MR application development using the HoloLens device. In the next chapter, we will focus on HoloLens as a device. Later, our focus will be on application development, taking you through the step-by-step process of developing a new holographic application and deploying it on the HoloLens device.

2
HoloLens – The Most Natural Way to Interact

Microsoft HoloLens is the most powerful MR personal computing Windows device. It allows you to place your digital content in the real physical world where it matters most to you. This is what is called a MR device, one that tries to blend the real and digital worlds seamlessly and provides new ways to visualize the world beyond the screens. The boundaries between the real and the digital world start to blur, and you can interact with the digital world very similarly to how you would interact with the real physical world.

What is it about HoloLens that makes it stand out from the other devices in this genre? It is the fact that Microsoft HoloLens is the first *fully untethered*, *self-contained* computer running on Windows 10. Yes, that's right: no hassle of wires, additional cables, or connection to a computer or additional device. It is hands-free, self-contained device with a built in battery.

Using HoloLens, you see virtual objects placed in the real world that are merged so seamlessly in the real world that it appears as if they are a part of it. You can interact with them using gestures as well as spoken commands, very close to how you would interact with real entities.

From this chapter onward, the focus will be on MR development using HoloLens. By the end of this book, you should be able to develop independent HoloLens applications. Another important focus of this book is to guide you through enterprise--integrated scenarios and how HoloLens applications can be developed to have external integrations and seamlessly work with enterprises' **line of business** (**LOB**) applications. In this chapter, we will cover the following topics:

- Components that build the HoloLens device
- Holograms and the real world
- Interacting with holograms

- Exploring different possibilities with HoloLens
- Key differentiator for the HoloLens device
- Understanding the development process
- Building up teams for HoloLens app development

Before we delve into the development process and start building projects, we need to have a good understanding of the device and the different types of application that can be developed using this device. To develop holographic applications using HoloLens, it is important for us to understand the different components it interacts with and the different ways to interact with virtual objects when they blend in real environment.

A closer look at HoloLens hardware

This self-contained, fully untethered holographic computer, the HoloLens, is a see-through combination of holographic lenses that capture data using an array of sensors and uses the power of processing units to understand the real and holographic environment around you. This hands-free device is made up of several sensors, cameras, and lenses put together in a lightweight head mounted device. It can be comfortably fitted over the head and adjusted as per your comfort.

It consists of specialized speakers, placed just above the ear position, which produce a floating surrounding sound. Everything is elegantly fitted and secured inside the HMD. The device produces data every millisecond and the power of processing units will take care of it with the ultimate design and processing powers.

To develop applications, it is essential to understand the different components of HoloLens because ultimately it is the hardware that will reflect the things that you want your application to do. The following are the key components of the HoloLens device:

- Cameras
- Sensors
- Lenses
- Speakers
- Microphones
- Processing Units
- Inertial Measurement Unit

Apart from the preceding components, the device also has a power adapter for an external power supply to charge the battery using a Micro USB Connector. The device has full support for connectivity with the outside world using built in Wi-Fi and Bluetooth; we will discuss about this in the subsequent sections.

The following figure shows the different high-level components and their placement positions inside the HoloLens device:

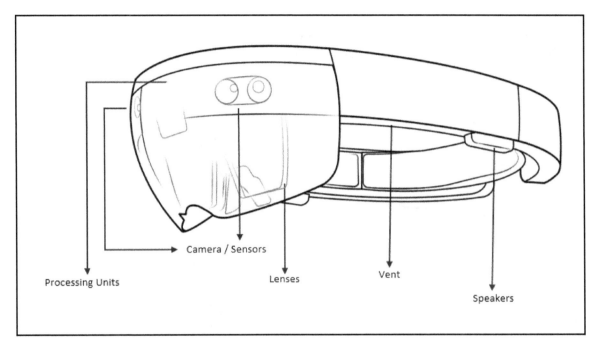

High-level components of HoloLens

Inside the HoloLens device

You can wear the device comfortably with the additional support of the headband that holds the device. The headband has an additional adjustment wheel to ensure that the device can fit different users comfortably. The magic starts when you put on the device and turn it on. You will start feeling the immersive experience in both the real and virtual worlds. You will be able to explore the virtual world in the real environment.

Having said that, nothing is possible without the combination of all internal components working together. Let's have a detailed look into the different HoloLens components and explore how they work together to redefine our personal computing through a new era of holographic experiences.

Cameras and sensors

HoloLens contains an array of sensors and cameras that accumulate data from the real environment. In the top-front part of the HoloLens, there are several front-facing cameras and sensors that are used to scan the environment and collect as much data as possible.

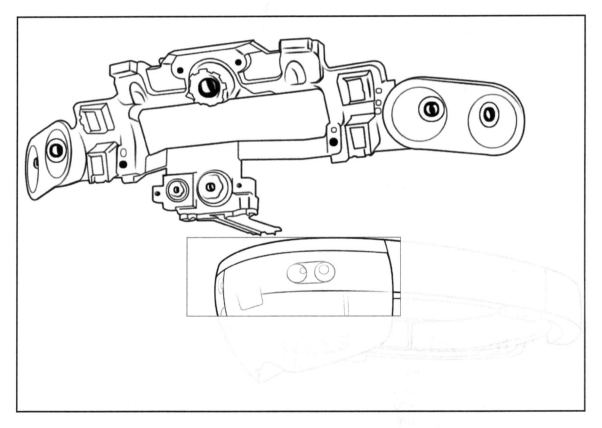

Sensor Bar placement in HoloLens Device

Sensors pass the data to processing units, which make it understandable by the HoloLens. At a high level, the sensor bar on the HoloLens embraces the following:

- Depth Camera
- Ambient Light Sensors
- Environment Understanding Camera
- Video Camera

The following figure illustrates the different types of sensors placed in the sensor bar:

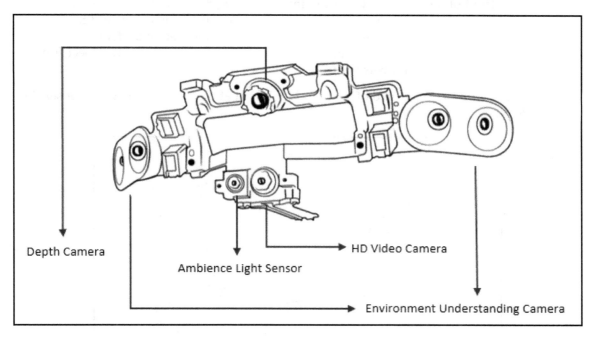

The Sensor Bar – Sensors and Camera

It all starts with the Depth Camera. The HoloLens Depth Camera is based on the Kinect technology, which uses a combination of camera and infrared depth-sensing vision to draw the digitalized surroundings by measuring the distances and depth.

Kinect has been a game changer in the world of motion games and applications. It is a motion sensing device, which was originally developed for the Xbox 360 gaming console. Kinect provides a **Natural User Interface** (**NUI**) for interaction using body motion and gestures as well as voice commands. It is a horizontal device with depth sensors, color camera, **infrared** (**IR**) Emitter, and a set of microphones, all secured inside a small box. This small box is attached to a small motor and tilts the device in a horizontal direction. When it is working, the IR Emitter constantly emits infrared light in a *pseudo-random dot* patterns over everything in front of the device. IR dots are invisible to us, but IR depth sensors can read those dots. The doted lights reflect from different object, and the IR depth sensor reads from the objects and converts them into depth information by measuring the distance between sensor and the object from where the IR dot was read.

HoloLens device has a wider field of view and looks out at the world with a field of sensing 120x120 degrees.

View area for HoloLens

Ambient Light Sensors maximize visual quality and energy conservation within the device, depending on the change in environment lights. An ambient light sensor is mounted in front of the HoloLens to detect the environment's light quality and intensity, which help the device in controlling the brightness. This helps to reduce power consumption, extend battery life, and provide optimum viewing in diverse lighting conditions.

The Environment Understanding Camera builds a quick understanding of the environment and provide the basis for head tracking and surface reconstruction. This is the key to understand your physical world and place holograms on physical objects.

HoloLens has a 2MP HD Video Camera that is used to record and share the visuals captured by the device and capture real-time photos. Camera feed can also be used to share the visual experience among different users.

The Lenses

Cameras helps the HoloLens see and understand the real environment, but the exciting part is how holograms are projected into your vision. HoloLens see through holographic lenses. We see the real world through the device's lenses, and holograms are projected out in front of us, up to several meters away. This gives us a mixed experiencing by blending both real and virtual worlds.

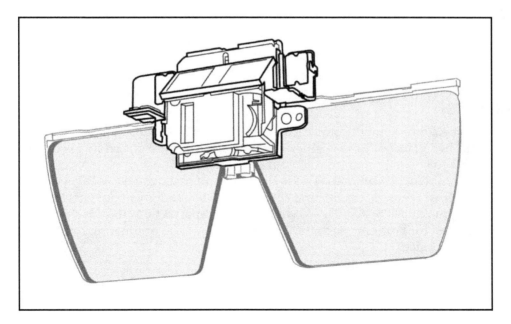

The Lenses

The HoloLens display is basically a set of transparent lenses placed just in front of the eyes. When light is emitted from the camera, it passes through the lenses; the light is reflected in your eyes. These lenses use an optical projection system to ray cast the holograms into your eyes as particles of light. Each lens lets light pass through and create different images to create a stereoscopic illusion of 3D.

Projection of holograms through Lenses

Speakers

Immersive experience without sound is incomplete. The HoloLens device has two built-in speakers attached just above the position of your ears. They are placed in a position such that they seem to be floating in front of your ears, and thus it doesn't block out the sound of the real world, unlike headphones. This is the beauty of having these speakers. The sounds produced by these speakers are floating surround sound, and give your ears the feel of hearing things in real 3D Space. This is what is called **Spatial Sounds**, which means that sound is adjusted for both ears such that sounds appear to be originating from a certain direction and distance.

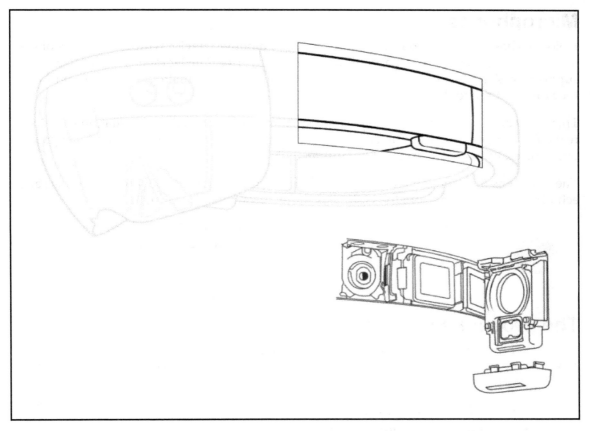

Speakers in HoloLens

There are volume buttons on the upper side of the device, which can be used to control the volume directly from the device itself.

Microphones

HoloLens has been designed as a hands-free head mounted device with voice as one of the key input mechanisms to accept commands from any user. The HoloLens device has a great support for Voice Commands using a set of microphones. The microphone array consists of four different microphones that take commands via voice control.

This array of microphones also helps in noise compression and echo cancellation from the real environment. The need for a microphone array is to clearly distinguish and recognize user Voice Commands over other environmental noises.

The HoloLens microphone array also uses beamforming to get a good, amplified noise and echo-free speech signal from the HoloLens for the user who is wearing it.

 Combining the different sound signals by identifying the sound source and listening to a particular direction is called **beamforming**.

The processing units

HoloLens processes millions of data items in milliseconds and, therefore, requires a lot of processing power. The wearable Windows 10 holographic device is built up with three "high end" processors as follows:

- **Central Processing Units (CPU)**
- **Graphical Processing Units (GPU)**
- **Holographic Processing Units (HPU)**

Although CPU and GPU are the most commonly used processing units used by all PCs, the most impressive part of HoloLens is its HPU0. The CPU (32-bit Intel-based), RAM, Wi-Fi chip, Bluetooth chip, and GPU are all on a tiny board and uniquely shaped for HoloLens. Understanding the real environment and rendering 3D holographic objects in the user view require a lot of processing power, and those are supplemented by the HPU.

Inertial Measurement Unit

The **Inertial Measurement Unit** (**IMU**) is an electronics module, which collects angular velocity, specific force, and linear acceleration data using a combination of accelerometers, gyroscopes, and magnetometers. In HoloLens, the IMU is mounted on the holographic lenses, right above the bridge of your nose.

Environment Understanding Cameras provide us digital space of our real world; the IMU captures the movement of our head and calculates the latest position information into the display as quickly as possible so that our device can render the information on field.

Vent

The HoloLens is powerful, more powerful than an average laptop. It continuously processes millions of data every millisecond. Despite such a high processing rate, it does not overheat. It has downward vents placed on both sides of the band, which maintain the airflow from the device. This keeps the device at a comfortable temperature range so that there is no discomfort even when it is worn for an extended period.

Connectivity

HoloLens has great support for connecting with external subsystems. The following are types of connectivity provided by the HoloLens:

- Wi-Fi
- Bluetooth
- Micro USB
- Audio Jack

HoloLens supports of both Wi-Fi and Bluetooth connectivity to connect with any external system over a network and by pairing. Connecting HoloLens to a Wi-Fi network or with a Bluetooth device is similar to connecting any other Windows 10 devices. HoloLens provides a easy and quick way to pair with Bluetooth device, and the most common Bluetooth accessories used with HoloLens are the HoloLens Clicker and Bluetooth keyboards.

The Micro USB is used to charge the device using the external power adapter and as well as to connect the HoloLens physically with any PC if you want to plugin and deploy your app directly.

Using the Audio Jack, you can plug in the headphones and the Microsoft HoloLens will seamlessly transfer the sound to the headphone.

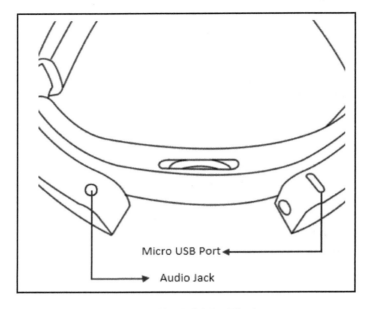

External Connectivity – Audio Jack and Microphone

Control buttons

While the Ambient Light Sensors help the device in controlling the brightness automatically, you can directly control the brightness of holograms as well as the volume from the HoloLens device using control buttons that are attached to it. Volume buttons are placed on top-left and brightness buttons are placed on top-right.

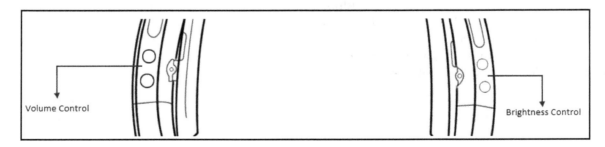

Volume and Brightness Control

LED indicator

Five dotted LED indicators are placed on the right-end side of the device just along with the power button. The indicator indicates the battery charge level and whether the device is on or off. Five illuminated LEDs indicate that the battery is fully charged. Along with the use of device, the decrease in the power level is indicated by the decrease in the illuminated LEDs.

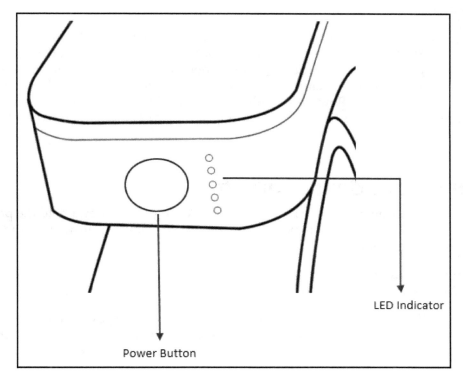

LED Indicator and Power Buttons

Putting things together

HoloLens is a fully untethered wearable device, running on Windows 10 with three built-in *high end* processors: a CPU, a GPU, and a special HPU. All the processing units, cameras, sensors, speakers, memory, Wi-Fi, and Bluetooth are intelligently placed within the device. A special vent works to keep it cool.

To summarize, holographic lenses are transparent so that you can see the real world behind the virtual world blended together. Holographic objects are the light particles, which bounce around millions of time to render the objects. Depth Camera uses infrared depth-sensing vision to draw the digitalized surroundings by measuring the distances and depth. It has a field of vision that spans 120 by 120 degrees. Sensors understand the real environment and digitalize the information captured from the environments. Sensors also track where we are looking and adjust the display accordingly. The sensors can also detect hand gestures.

Processing units are the heart of the HoloLens. It contains the CPU, GPU, and HPU, which process terabytes of data every second. The two built-in speakers attached to the device product floating sounds and an array of four microphones is used to capture the user's voice clearly irrespective of the environment. The device is more powerful than a laptop. The vent maintains cool air flows to the sides to keep the device within the normal temperature. Buttons on the sides allow us to adjust volume and control the contrast in the hologram.

Turning your device on/off

There is a power button next to the LED indicators. When you press and hold the power button for a second, the device will be turned on. When the device is turned on, the battery status LEDs will light up.

To turn on the device, press and hold the power button just for a few more seconds, probably 3-4 seconds. The device will be turned off when the LEDs fades.

 When we connect the device with PCs or any other power source over USB, HoloLens will turn on automatically. HoloLens will also turn off automatically when the remaining battery power drops below 2%.

Cortana on HoloLens

Cortana is your personal assistant working on your phone, PC, and Tablets running on Windows 10. The HoloLens device running on Windows 10 is no exception; it has support for Cortana.

By default, Cortana is on when you use the device for the very first time. You can start by just saying "*Hey Cortana*" anytime while using the HoloLens. Be it opening an App, controlling the brightness, taking a picture, or even shutting down or restarting your device, you can always use Cortana's assistance.

At the time of writing this book, Cortana on HoloLens was available in English only.

You can turn Cortana off or change its settings from Cortana's settings options.

To shut down your device, you can just say "*Hey Cortana, turn off*". Cortana will ask for re-confirmation, "*Are you sure you want to shut down?*" If you reply, "*Yes.*", the device will be turned off.

The Mixed Reality world - HoloLens shell

HoloLens shell is the MR world that contrives the immersive experience with holograms and the real world. Whatever we see through the HoloLens appears through the HoloLens shell. At the beginning, the shell launches with the start screen, similar to the one for any system running on Windows 10.

The start screen is the first point of interaction with other holograms. To launch a holographic app, open up the **Settings** options for HoloLens.

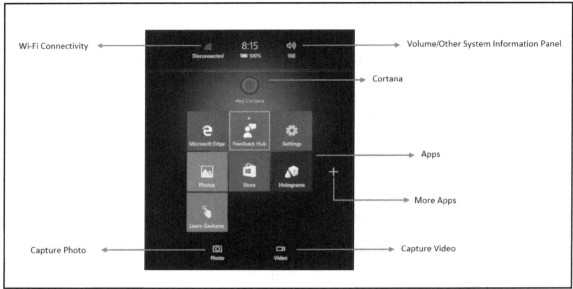

HoloLens start menu

The Start menu of HoloLens consists of basic system information, including time, Wi-Fi connection status, battery status, and many more. All this information is pinned up on the front view along with all other apps, which we can navigate through. It also consists of buttons to capture photos and MR video along with options to invoke.

Holograms in reality

Till now, we have mentioned Hologram several times. It is evident that these are crucial for HoloLens and Holographic apps, but what is a Hologram?

Holograms are the virtual objects which will be made up with light and sound and blend with the real world to give us an immersive MR experience with both real and virtual worlds. In other words, a Hologram is an object like any other real-world object; the only difference is that it is made up of light rather than matter.

 The technology behind making holograms is known as **Holography**.

The following figure represent two holographic objects placed on the top of a real-size table and gives the experience of placing a real object on a real surface:

Holograms objects in real environment

HoloLens, holograms and your real world

Holograms are physically intangible. We can't touch them; however, we can view them from a distance and from different angles. Different angles give us a different perspective of the same holographic object just like how it works for a real object.

Using HoloLens, we can interact with holograms virtually, which internally applies the transformation to those light objects visible in the physical environment and we can see those objects changing. Holograms can be both 2D and 3D, depending on the type of object created, and it persists the position where it has been adjusted though the HoloLens. It will remain in the same position even if you leave the environment and come back or even if you restart your device.

 Holograms will remain persisted in the same position unless you move them into other positions; this is known as Hologram persistence.

Hologram and the Windows Holographic platform

Windows Holographic is the MR platform for developers built on the Windows 10 API. The developer can leverage the power of **Holographic API** with Windows 10 and build immersive experience apps using HoloLens.

Recording holograms – Mixed Reality capture

Using HoloLens, *you can record what you are seeing through the HoloLens*. We have already seen that HoloLens has a *HD Video Camera* that is used to record and share the visuals captured by the device. HoloLens' cameras allow apps to capture what the front-facing camera sees (real environment) and the holograms that you're seeing.

To start capturing either **Photo** or **Video**, you can just launch the photo/video app from the Start menu:

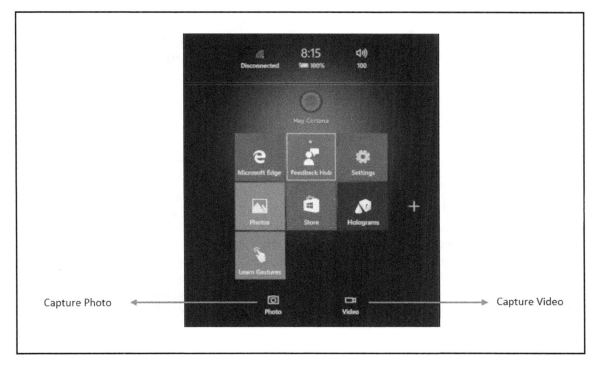

HoloLens start menu

Alternatively, you can ask Cortana to capture a photo or start recording video. There are other alterative options to capture MR video or photos from the Device Portal, which we will cover in a later part of this chapter.

 During the recording, the HoloLens frame refresh rate will drop.

HoloLens interaction and application model

So far, we have learned about the HoloLens device and the holograms in the MR world. HoloLens projects holograms and blend them into the real world to give an immersive experience using both virtual and real objects. What next? Now is the time to explore how to interact with these holograms to feel them as real as your real objects are. What are the different ways to interact with them?

In this section, we will cover the different ways of interacting with the holograms and the application life cycle for holographic applications.

Interacting with holograms

There are basically five ways that you can interact with holograms and HoloLens. Using your *Gaze*, *Gesture*, and *Voice* and with *Spatial Audio* and *Spatial Mapping*. Spatial Mapping provides a detailed illustration of the real-world surfaces in the environment around HoloLens. This allows developers to understand the digitalized real environments and mix holograms into the world around you. Gaze is the most usual and common one, and we start the interaction with it. At any time, HoloLens would know what you are looking at using Gaze. Based on that, the device can take further decisions on the gesture and voice that should be targeted. Spatial Audio is the sound coming out from HoloLens and we use spatial audio to inflate the MR experience beyond the visual.

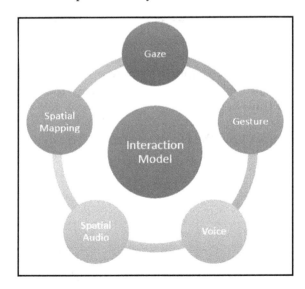

HoloLens Interaction Model

So, here are the five key elements for HoloLens interaction:

- **Gaze**: Indicates where the user is currently focusing
- **Gesture**: A user action performed on any of the gazed items
- **Voice**: Executing commands based on your Voice Commands
- **Spatial Mapping**: Understands your real environment and helps mix your holograms to the real surface
- **Spatial Audio**: Spatial Audio to expand the MR experience beyond our visual senses

Before we go into more detail on all these interaction models, we can consider Gaze, Gesture, and Voice, to make it more logical, as input for HoloLens from users, whereas Spatial Mapping takes inputs from the environments to understand your real environment. On the other hand, Spatial Audio can be considered as an output interaction for HoloLens.

Let's explore them in detail.

Gaze

In our PC/laptop, we use a mouse to move a cursor. **Gaze** is the mouse for the HoloLens. Gaze is one form of input for HoloLens, which *indicates where the user is currently looking*. Similar to the mouse cursor, in Holographic apps as well we use an indicator to show the direction of the Gaze. The cursor follows the user's head moving in real space.

To determine the point where the user is currently focusing on, each frame does a raycast into the physical space and places the cursor on the object where the raycast intersects with the objects. Even if the user is moving, HoloLens still raycasts for each frame and thus maintains the focus. All the data for the raycast is captured by the sensors and IMU helps in tracking the user movement.

The following figure show how the Gaze object changes based on the user's head movement from one to another object.

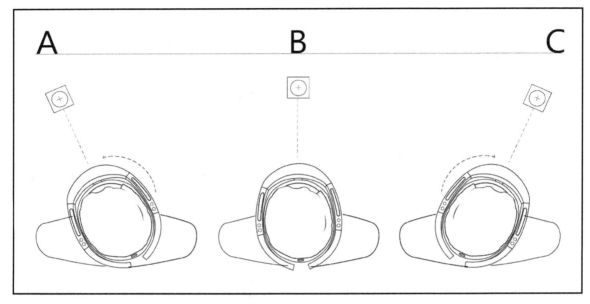

Gaze in HoloLens

The cursor indicates the Gaze direction for any holographic application and can also indicate which item can be gazed and which cannot.

Gesture

Gesture works in the combination with Gaze, or followed by Gaze. First, Gaze on some object, then take an action using Gesture. While Gaze is like moving the mouse cursor, Gesture in HoloLens is like a mouse click or tapping in a touch-based application.

The sensors inside HoloLens can recognize a hand gesture as long as we keep our hands inside the viewable range of the HoloLens.

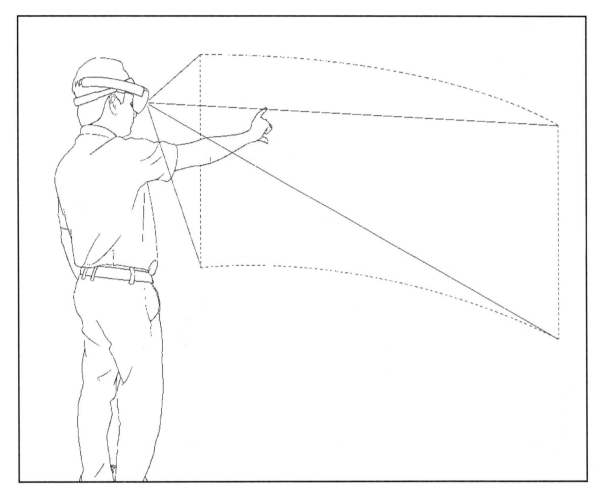

HoloLens viewable range and gestate action within the viewable range

There are several types of gesture, including the following:

- Bloom
- Air Tap
- Tap and Hold

The **Bloom** gesture will invoke the Start menu; we discussed about it in the HoloLens Shell section; blooming again will dismiss it. So, at any point , if you want the start screen back while using HoloLens, you can simply use the bloom gesture. Using the Air Tap, you can select holograms and take action on them. Air Tap is like a mouse click, where we can take an action by tapping our finger down and then taking it back up. Tap and Hold works like drag and drop with a mouse. Using this gesture experience, you can move an object from one position to another position.

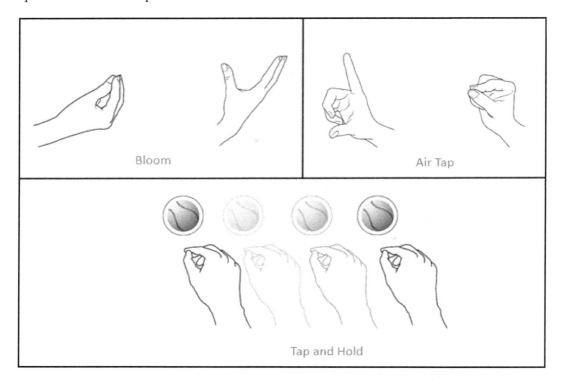

Illustration of Bloom, Air Tap and Tap and Hold Gesture

Voice

Voice is another form of input on HoloLens on the Windows Holographic platform. Once you've targeted a hologram with your gaze, you can interact with that hologram by giving a voice command. The Voice Commands are very convenient for a hands-free device, such as HoloLens. You can walk around the room and just speak out what you want.

Voice Commands can be categorized into the following types:

- System Voice Commands
- Custom Voice Commands

System Voice Commands are fixed and always available. The following is a list of system Voice Commands:

- Select
- Place
- Face me
- Bigger/Smaller
- Hey Cortana, go to start
- Hey Cortana, shut down
- Hey Cortana, move <app name> here

As a developer, you can define your own Voice Commands and actions. Voice commands are implemented based on standard **Universal Windows Platform (UWP)** voice command patterns and the use of the Windows Holographic platform. Having implemented a custom voice command, you must have to listen for the event handler for the same Voice Commands, very similar to how general event handling works in our development.

HoloLens has the inbuilt Cortana; you can take it to the next level by integrating your Voice Commands with Cortana.

Spatial Audio

Spatial Audio expands the HoloLens MR experience beyond the visual senses. By taking advantages of the spatial audio, a developer can greatly experiment the MR world of HoloLens. Holograms are virtual objects; they can appear from any direction in the scene whereas HoloLens has only field of view in the front. If you are looking at your front, and there are some holograms that are coming toward you from the back, then a floating sound from the back can add to the experience and make it more real. With Spatial Sound, sounds can come from any direction and create a 360 degree of holograms representation.

HoloLens uses **head related transfer function (HRTF)** to bring the 3D floating sound using the direction and distance of the objects. This makes MR environments more realistic. Spatial Audio capabilities with HoloLens provide the developer with a lot of capabilities.

Spatial Sound greatly enhances the HoloLens user experience by making our imaginary world more complete.

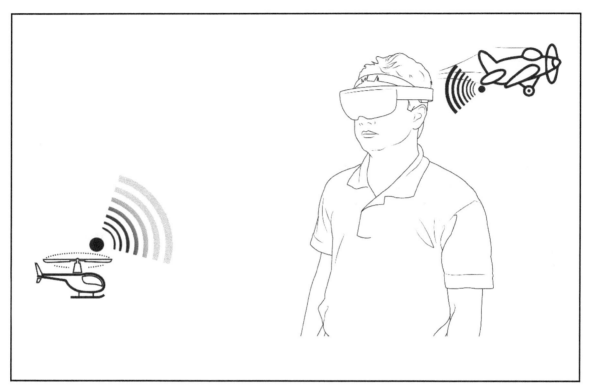

Illustration of Spatial Sound from different distances

Spatial Mapping

Spatial Mapping is key for the holographic experience. It provides a detailed representation of real-world surfaces in the environment around HoloLens. Spatial Mapping collects all the information about the real world to use in the holographic world and allows developers to mix holograms into the world around you. The following figure illustrates a Spatial Mapping view from a HoloLens device:

Spatial mapping view of a real space

We use Spatial Mapping to understand the environment and where real objects are. This helps the placement, navigation, and visualization of any holograms in the real space.

As we proceed toward development in upcoming chapters, we will explore all these interaction models in detail with real implementations.

HoloLens app life cycles

All Holographic apps running on HoloLens are UWP application. The following illustration represents the app model state for an app running on Windows Universal Platform. If you are familiar with developing Windows Universal app, you must be familiar with this app state model. Till now, you may have experience in developing 2D Windows Universal app for desktops, tablets, or phones. Windows Universal app for HoloLens is one step ahead; now, you can develop not only 2D apps, but also 3D apps. In this book, we will focus only on 3D application development to realize the power of MR.

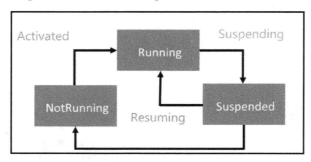

Application state flow

The following are the high-level steps for a Holographic app life cycle:

- Placement
- Launch
- Execute
- Terminate
- Remove

App Launch starts as soon as we place the app into the HoloLens Shell. Once the app is launched, we can change the placement of the app within the HoloLens Shell. Execution takes place once the app launch is completed. During the execution phase, the app goes through the app life cycle model of Resuming–Suspending–Activated state. The app life cycle is completed by removing the app from the HoloLens Shell.

The following diagram shows the application life cycle flow for a Holographic app running on HoloLens:

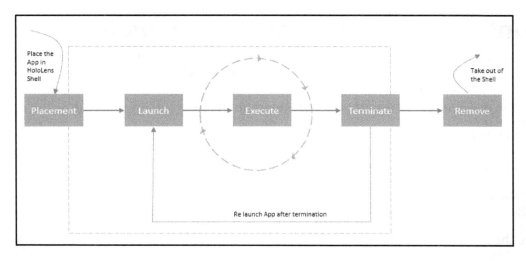

Holographic app life cycle

HoloLens Clicker

HoloLens Clicker is an accessory that comes along with HoloLens. It connects to the HoloLens using Bluetooth and allows a user to click and scroll with minimal hand motion. You can consider this as a replacement for Air Tap gestures where, instead of using hand gestures, you can use the Clicker. When an object is gazed, you can press the button on the Clicker, which will execute the Air Tap command for the selected object.

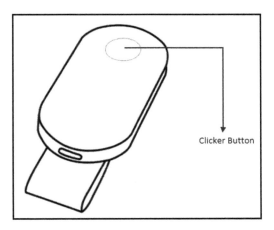

HoloLens Clicker

Controlling your HoloLens remotely

The HoloLens Device Portal lets you manage and configure your device remotely over a network. You can connect, deploy apps, and capture and view real-time MR data using the HoloLens Device Portal. It has also advanced diagnostic tools to help you troubleshoot and view the real-time performance of your Holographic app.

The HoloLens Device Portal is like the Windows Device Portal and is available for each device family. It is a web server on your device that you can connect to from a web browser on your PC when running on the same network.

To enable access to the device portal, you must enable the Developer Mode by turning on the Remote Management access from the **Settings | Updates & Security | For Developer**.

Once the device setup is done, you can access the device portal in your computer browser just using your HoloLens device IP address. The following screenshot shows a first glimpse of your HoloLens Device Portal:

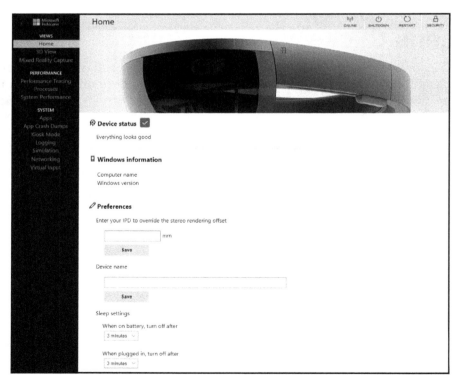

HoloLens Device Portal

The Device Portal has several interesting features including network simulation and virtual input acceptance.

Key differentiators

HoloLens is first a MR device; let's see how it differentiates itself from other AR/VR experiences available in the market.

Immersive with surroundings

In a typical VR device, you will be completely cut off from the surroundings; that means you will not know what's going on right in front of you. Microsoft HoloLens is different in that respect; it doesn't cut you off from the surroundings. After wearing a HoloLens device, you will continue to see, hear, and respond to whatever is going on around you.

Independent standalone device

HoloLens is a completely independent device--it is not dependent on any PC nor on any smartphone. It's not like any other VR or MR devices, which require physical connectivity to either a PC or smartphone. Due to this feature, it's very versatile and mobile in nature; its user can carry it anywhere like another mobile device.

No screen staring

Unlike a typical VR, where you are forced to stare at a screen in front of you, within HoloLens you still have a lens in front of you, but it's a see-through lens. That means you can see holograms that are generated on that screen along with the real surround objects, and it interacts with reality as well as with holograms.

Mixed not virtual

While VR devices take you completely within a virtual 3D World/Space, HoloLens brings the virtual world within the real world, and allows the user to interact with those virtual objects. Even the interactivity is quite simple in nature, through Voice Commands or Air Taps similar to mouse clicks.

Broad scope

While VR devices are mostly used for gaming and entertainment, HoloLens is already being used for a varied set of implementations. NASA is using HoloLens to deliver a real-life Mars experience within the room. Volvo is using it for demonstrating its upcoming cars to users. Users can do their day to day work along with HoloLens using Skype chat and provide remote help. HoloLens is also used in the field of education and health.

Exploring the possibilities with HoloLens

The way MR delivers unique experiences, companies and enterprises have already started exploring a variety of scenarios for its applicability. Let's explore some of these possibilities:

- Virtual collaboration
- Remote expert guidance
- People-centric application

Virtual collaboration

MR gives an opportunity to collaborate using virtual object in real space. Team members can work together on the same or shared virtual objects in real space, brainstorm ideas, and collaborate in an altogether different way. This collaboration scenario can be used in many different enterprise use cases. One of them could be sales representative who collaborate with a customer and explains to them about a new project through virtual objects, and the customer can visualize the project in 3D Space.

In the building/construction industry, architects/designers can use MR devices to share the vision of the ongoing construction and even collaborate on ideas about its development.

Remote expert guidance

Nowadays, the workforce of an enterprise or a company is not located in one place, but spread out across the world. With such a distributed workforce, providing training and guidance across the globe is a challenge. People can use conference calls and video calls, to collaborate, but both have their own limitations.

That is where MR brings the hands-free experience of interaction, where remote expertise can help local resource by collaborating using MR devices.

Some of these scenarios where MR devices can be used for providing remote guidance could be aircraft maintenance, oil refinery maintenance, healthcare, and many more.

People-centric applications

Till now applications within enterprises have been created to represent data in 2D formats. With MR, the same data can be represented in a 3D Space, while intermingled with the respective/relative real objects.

For example, managers used to review 2D reports to view an update on the manufacturing in a factory. Now, they can wear an MR device, walk through the manufacturing unit, interact with employees, and updates can be virtually overlaid over the respective machinery or sections.

Another application could be during war, where soldiers can collaborate and share data in real-time among themselves.

MR devices could also be used as a navigation device while driving. Hence, there is no need to get distracted by smartphone or GPS devices; the MR device can project navigation directions right into the view of the driver.

Well, these are just a few special examples of domains where you can use the HoloLens. However, at the end of the day, it's up to your imagination where and how you want this device to work.

Development process and team building

We learned about the HoloLens hardware, application, and the interaction model required for application development. Now, let's learn about the type of skill set required for designing and developing MR apps using HoloLens.

Building up the team

A HoloLens development team is like any 3D game development team. It mainly consists of designers, 3D Artists, developers, audio, and video designers.

Designers

A designer is expected to have experience in running and building prototypes, and managing the quality of the experience throughout production, as well as pre-production. Designer responsibilities include maintaining quality throughout experience, building out experience, design and build prototypes, documenting, and presenting the experience to clients.

3D Artists

The 3D Artist is responsible for creating high-quality and the performance 3D digital models, plus their textures and surfaces, in a wide range of types and styles; it includes realistic hard-surface models of architecture, props, vehicles, machinery, and other objects, as well as organic models of landscape elements, scenery, and characters. Ability to work low-fi (low fidelity) to high-fi (high fidelity), use reference materials or work from existing plans or designs and understand proper scale and proportion.

Developers

The developer is responsible for developing both user-facing and technical features in HoloLens applications. Responsibilities include prototyping and the development of user interaction mechanics, overall application flow, and key technical features.

Audio designers

The audio designer is tasked with authoring audio content for HoloLens utilizing HRTF and other platform-specific audio development tools and implementation standards. The audio designer works creatively to establish audio targets as part of the various design directives and explores creative opportunities to exploit the unique sonic capabilities of HoloLens.

Leads

The lead is responsible for app creation from concept to completion. Responsibilities include creating and driving to the schedule, planning and costing for all tasks, task prioritization, driving completion of work from all disciplines to high quality, client communications, and documentation.

Development process overview

The development process for HoloLens application is like any other 3D game development. Any application development process goes through different stages of development, and the following is one example of it. It's not like there is a hard and fast rule to follow this process; it is just a guideline, and you can tweak it as per your requirements and process. At a high level, the development process goes through the following six stages of development:

- **Requirements Envisioning**
- **Storyboarding**
- **Planning, Design and Prototyping**
- **Development**
- **Deploy and Test**

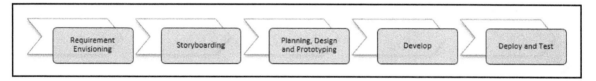

Phases of development

Requirement Envisioning

An objective of this phase is to identify the primary scenario for the HoloLens App development. The customer's team and the designer meet during this phase and brainstorm ideas related to the customer's requirements, and where HoloLens can play a role. Discussions start from ideas about new possibilities with MR. It could be around remote assistance, virtual collaboration, or people-centric applications, but in the end, the target application should make the customer's team more productive.

This brainstorming can take up to a few days, as each idea is scrutinized from the angle of possibility, feasibility, and productivity. The final output of this activity is a single priority idea, which will be taken up by the team for further development.

Storyboarding

Storyboarding involves outlining and sketching primary flows of the experience or scenarios. Designers will sketch scenarios, show interactions through Gaze and Gestures, and revalidate them with their customer.

The following are a few examples showing interactions with a virtual car using gestures:

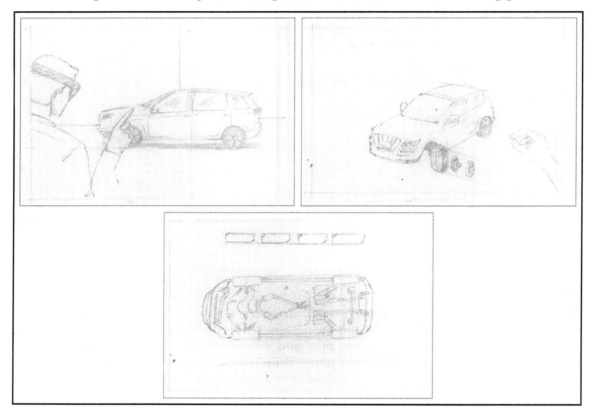

Planning, Designing and Prototyping

After completing the Requirement Envisioning Phase, you move to the next phase, Planning/Design/Prototyping, where you will detail out the storyboard, define specifications for 3D Objects, and detail out the interaction model.

Planning and Design

As the part of Planning and Design activity, you will detail out storyboard with all the information about navigation flows, scene actors, and objects; required animations; and the size and positions of objects. This can be achieved through a series of design meetings among the artist, designer, and developers. The following is a set of activities, which is generally followed:

1. Identify each scene and break it down into required objects, user, information/text, audio, and animations.
2. Object sizing and positioning; identify the size and position of primary and secondary objects within the scene.
3. State of objects; define the state of objects when they are gazed, un-gazed (normal state), and Air Tapped. This is basically defining the transformation of objects based on user actions.
4. Actions/Gestures define interactivity within the scene, using gaze, gestures, and Air Taps. Using these actions/gestures, you will basically define the flow in the story.
5. UI/UX defines the size, color, and font for every object/text within the scene.
6. Narration defines speech or audio interaction within the scene and also commands or narrations to be used within the scene.

Prototyping

The prototyping process consists of various small **proof of concept** (**PoC**) activities to validate design decisions, such as best size, position, animations, interactivity, and so on for or within scene components. The main purpose of these prototypes is to try out several things on the scene components and then finally come up with the best experience.

Develop

Once the design and prototyping are completed, the actual development process starts with each scene being developed using prototypes. The development of assets and scripts is done in parallel. 3D Assets are being developed by 3D Artists, and scripts and flows are being developed by developers.

First, the initial assets are provided to developers, animation and flow scripts, and then detailed assets are created in parallel.

The designer's role is to continuously review and provide feedback to both 3D Artists and developers, and keep the overall development in sync with the original vision of the customer.

Deploy and Test

After the development is completed, you deploy the application on HoloLens and share it with your customer for feedback. This exercise can also happen in parallel to development phase if your team is using the iterative development model, that is, you can share interim builds to the customer for early feedback after you have completed each iteration cycle.

There are several ways of deploying the application, and it depends on what stage of your application you are in and the intended user of your application. During the development phase, you can deploy it using Visual Studio, Device Portal, or even using the Unity Remote deployment tool. We will cover them in detail in subsequent chapters.

In the case of releasing the application for end users, you can choose any of the following method to deploy:

- Through Windows Store app deployment
- Through side loading deployment

Both these approaches of deployment are very common for any UWP, and don't forget, all Holographic apps are UWP app only. If you want to distribute apps to abroad set of users both inside and outside your organization, you may do so through Windows Store app deployment, which will take you through Windows Store certification process. In this case, you need to target your store deployment for the Holographic platform.

For a limited set of users, or for an internal enterprise application, you can go with the *Side Loaded* deployment process. Now, here you can use several processes to side-load the packages. One of the common steps would be to create the package and push it into the device using Device Portal. We will cover more on these topics in the following chapters.

Summary

This chapter provided you with an inside look at the different components of the HoloLens device. You saw the major components of the device such as camera, sensors, lenses, speakers, and a set of processing units. The camera and sensors ensure the capturing and projecting of holograms, which is of prime importance for the functioning of the device; the speakers, on other hand, ensure the floating and surrounding audio that give a mixed experience along with the 3D view objects. It is also worth mentioning how different sets of processing units come together and process high volumes of data to produce an immersive experience. It would not be wrong to say that it is this combination of technological innovations that makes the HoloLens the awe-inspiring device that it is.

You have gone through the different possibilities of applications that can be developed using HoloLens. In the next chapter, we will jump-start building projects from scratch. Through these projects, you will learn not just about 3D application development for HoloLens, but will also learn about integration scenarios, which will be one of the core requirements for enterprises. You will learn to integrate with Azure services from within the HoloLens application, and dynamically consume data from cloud and change the user experience in real time.

3
Explore HoloLens as Hologram - Scenario Identification and Sketching

In the previous chapter, you learned about HoloLens and holographic apps and about the different components that build up the HoloLens device and how they work together. You also learned about the concept of holograms, their interaction with the real world, the application development process, application models and interaction models required for developing holographic application, and lastly, about the types of applications and possibilities with the HoloLens device. Let's start applying our learning from the previous chapter by developing our first holographic application and deploying it on HoloLens. Rather than jumping directly to development and coding, let's go through the complete journey of holographic application development. This journey starts from identifying a new scenario or application requirements, detailing out that scenario through a sketch and plan phase, creating 3D Assets, developing a scene, applying scripts, and finally, deploying it on the device and testing it out. This journey of holographic application development is divided into two chapters, this current one and the next one. In this chapter, you will learn about the following:

- Envisioning
- Scenario prioritization
- Sketching
- Asset design and development
- Things to consider while 3D using the Asset Pipeline
- Setting up a development environment
- Using HoloLens emulator

In the following chapter, you will continue this same scenario and do the development and deployment on the HoloLens device.

Envisioning

To make the whole development process real, let's assume you are working for XYZ Inc., which has a global presence and has employees spread out across the world at different office locations. XYZ has procured HoloLens devices and has also got the holographic application developed. Now, their business problem is to train their employees about usage and handling of HoloLens devices so that employees can start using these devices in their day-to-day work and be more productive.

To achieve this, the first thing you will do is schedule a meeting with XYZ leadership and stake holders, and learn about their vision and current problems. You can take different approaches to conducting such meetings with high-level stake holders, and one of the approaches could be to brainstorm ideas using sticky notes session. The following diagram illustrates this:

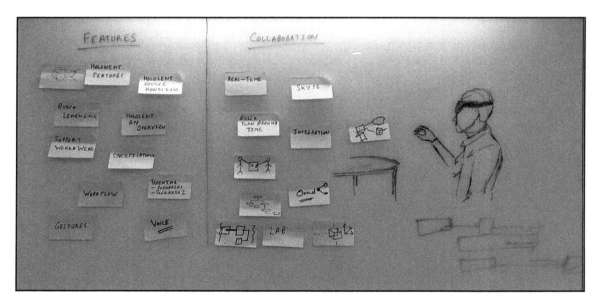

Sticky notes discussion and brainstorming session

Training scenario for HoloLens

From this meeting, one of the tasks, which came out collaboratively from all stakeholders, is to leverage HoloLens itself to train their employees so that they don't have to hire and send their trainers across the world. This will result in a direct saving of money during an employee's training. Different possibilities are discussed during that meeting to leverage HoloLens as a training tool. The following is a list of two primary scenarios, which were shortlisted within that meeting:

- **Features and usage:** Educate employees about HoloLens features and capabilities, and things they need to take care of while using it
- **Collaboration:** Educate employees about the usage of HoloLens tools for collaboration among themselves, such as using Skype on HoloLens

Features

The requirement is that employees should be able to learn about the following features and capabilities of HoloLens as a device:

- Learn about lenses and how holograms are forms and created using lenses
- Learn about speakers
- Learn about buttons to adjust speaker sound and brightness
- Learn about safety tips to keep HoloLens clean and safe

Collaboration

Employees should be able to leverage HoloLens for remote collaboration, as follows:

- Using Skype on HoloLens to discuss and share ideas
- Using Skype on HoloLens for remote collaboration scenarios, where a person can remotely guide someone wearing HoloLens

Scenario prioritization

After identifying scenarios, the next exercise is to prioritize scenarios. Scenario prioritized scenarios will be taken up for the next phase of the development. A team can either do the prioritization exercise in the same meeting or in another follow-up meeting. So, let's go through this prioritization exercise. During the prioritization exercise, the team discusses these scenarios from the perspective of immediate need and dependencies. After reviewing both identified scenarios of *Features* and *Collaboration*, they list down the following two factors to be considered in the prioritization exercise:

- **Immediate need**: Device handling to start immediately, as devices are already being procured
- **Dependency**: Collaboration can start only after employees are trained about the handling of the device

So, looking at immediate need and dependency, the team decided to prioritize the *Features* scenario over the *Collaboration* scenario.

The next exercise is to elaborate the prioritize scenario and start defining interaction details within this scenario.

Scenario elaboration

After the scenario prioritization exercise, the next activity is to elaborate the selected scenario. In this, you will detail out features and user interactions within the scenario.

The objective of the selected scenario is to provide the customer's employees with the readiness to handle the HoloLens device, as well as to educate them about hardware details. To elaborate the scenario, let's break it down into the following sections.

Primary View

It helps the user to view the hologram of the HoloLens device and roam around and view it from different angles. At the launch of the application, audio narration guides the user about the application's features. The narrative mentions everything they can interact with within this application and how to interact.

Lenses

When a user views the Lenses, or gazes at them, they should be highlighted, giving the impression that they are interactive. When highlighted, the user can Air Tap and the application should explain the following details about the lenses:

- Explain the role of lenses in HoloLens
- Explain the hygiene and cleanliness requirements for lenses

Speakers

When a user views the speakers, or gazes at them, they should be highlighted, giving the impression to the user that they are interactive. When highlighted, the user can Air Tap and the application should explain the details about the speakers.

Cameras

When a user views any of the cameras, or gazes at them, they should be highlighted, giving the impression to the user that they are interactive. When highlighted, the user can Air Tap and the application should explain different camera details and their purpose.

Volume control buttons

When a user views speaker volume buttons, or gazes at them, they should be highlighted, giving the impression to the user that they are interactive. When highlighted, the user can Air Tap and the application should explain how to use those buttons and adjust the volume control.

Brightness control buttons

When a user views brightness control buttons, or gazes at them, they should be highlighted, giving the impression to the user that they are interactive. When highlighted, the user can Air Tap and the application should explain how to use those buttons and adjust the brightness of holograms.

Sketching the scenarios

The next step after elaborating scenario details is to come up with sketches for this scenario. There is a twofold purpose for sketching; first, it will be input to the next phase of asset development for the 3D Artist, as well as helping to validate requirements from the customer, so there are no surprises at the time of delivery.

For sketching, either the designer can take it up on their own and build sketches, or they can take help from the 3D Artist. Let's start with the sketch for the primary view of the scenario, where the user is viewing the HoloLens's hologram:

- Roam around the hologram to view it from different angles
- Gaze at different interactive components

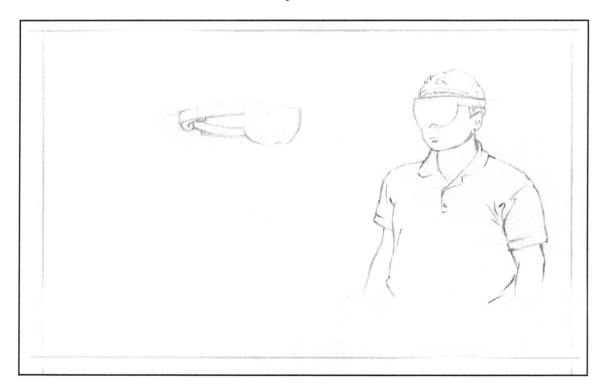

Sketch for user viewing hologram for the HoloLens

Sketching - interaction with speakers

While viewing the hologram, a user can gaze at different interactive components. One such component, identified earlier, is the speaker. At the time of gazing at the speaker, it should be highlighted and the user can then Air Tap at it. The Air Tap action should expand the speaker hologram and the user should be able to view the speaker component in detail.

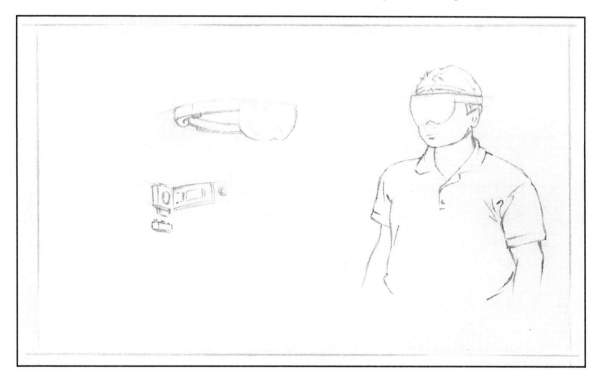

Sketch for expanded speakers

After the speakers are expanded, the user should be able to visualize the speaker components in detail. Now, if the user Air Taps on the expanded speakers, the application should do the following:

- Open the textual detail component about the speakers; the user can read the content and learn about the speakers in detail
- Start voice narration, detailing speaker details
- The user can also Air Tap on the expanded speaker component, and this action should close the expanded speaker

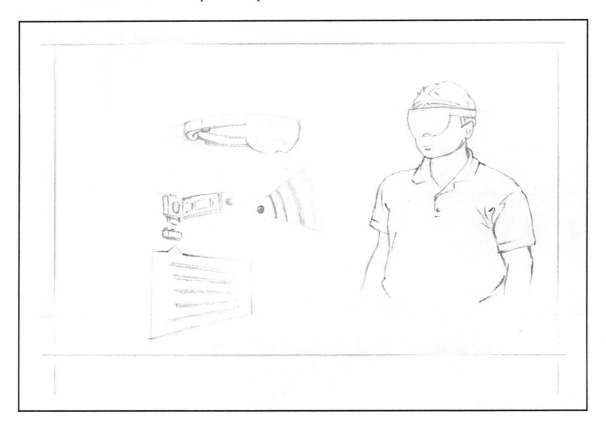

Textual and voice narration for speaker details

As you did sketching for the speakers, apply a similar approach and do sketching for other components, such as lenses, buttons, and so on.

Assets design and development

In the preceding section, you identified scenarios and detailed them out through the sketching process. The next step is to develop the 3D Assets that you will use in the next chapter to develop your first holographic application. In this section, we will provide you with an overview of tools which are generally used for 3D Asset development. For project continuation, you have a choice:

If you have any of these 3D Modeling tools, then you can try to create your own asset.

If not, then download the pre-built 3D Asset using the following link so that you can continue with the development of your first holographic application in the next chapter; refer to `https://github.com/PacktPubli shing/HoloLens-Blueprints`.

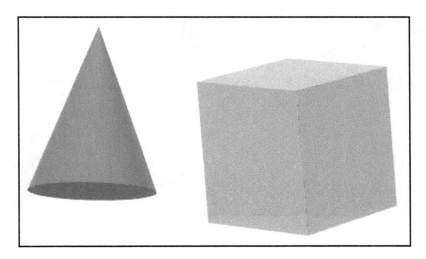

3D Models example for cone and cube

3D Modeling

3D Modeling is all about developing a three-dimensional object by recreating any real world object or any conceptual art. This can be achieved through different tools, and the process of creating three-dimensional objects using software is called **3D Modeling**. Industries such as video gaming, visual effect creation in motion pictures or animated movies, and VR games have been using this process of 3D Modeling for decades.

The basic approach of developing a 3D Model is to use a set of two-dimensional polygons and place them together in a three-dimensional space. A polygon is a two-dimensional object with a minimum of three-vertices, such as a triangle, square, pentagon, hexagon, and so on. The following is an example of a polygon in a triangle shape, where you can view the object in a three-dimensional space:

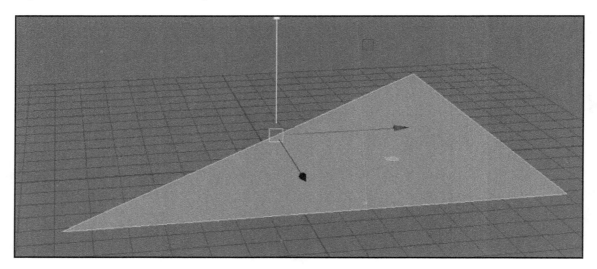

Polygon with three-vertices

Now, take this same polygon and add two more polygons in different directions/axis, and this will create a three-dimensional object.

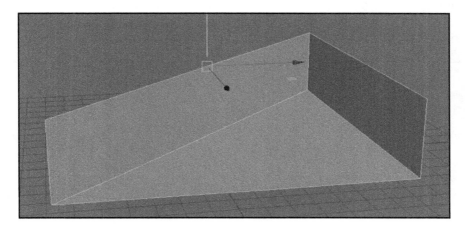

3D Model created with polygons

Further in this section, we will apply the same logic of creating three-dimensional objects using a set of polygons and 3D basic primitives, such as a cube or sphere.

3D Modeling tools

There are many 3D Modeling tools in the industry. Selecting a tool depends on the style of the modeling process used by the industry or the artist's choice. The following is a list of the most popular ones:

- Maya
- 3ds Max
- Cinema 4D
- Modo
- Mudbox
- Blender
- Lighwave 3D
- SketchUp
- ZBrush

In this book, we will use Autodesk Maya, Adobe Photoshop, and Unity3D for creating 3D Assets.

3D Modeling workflow

Before jumping to 3D Modeling, let's understand the 3D Modeling workflow across different tools that we are going to use during the course of this book. The following diagram explains the flow of the 3D Modeling workflow:

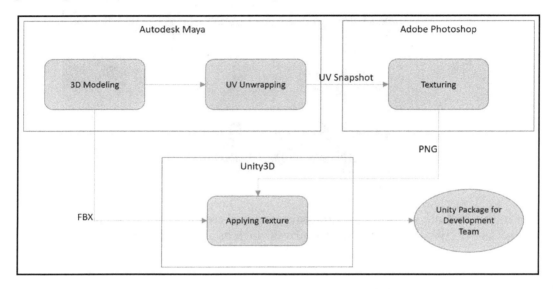

Flow of 3D Modeling workflow

In the next section, you will follow this workflow; first, you will learn to do **3D Modeling** and **UV Unwrapping** in **Autodesk Maya**. **Adobe Photoshop** can then be used to do **Texturing** on the UV Unwrapping output. Then, both outputs can be taken from **Autodesk Maya** and **Adobe Photoshop** to **Unity3D**. Using **Unity3D**, you can apply texture on them and publish the assets as a package to be consumed by the development team.

Build your 3D Model

In this section, you will learn to model a basic 3D Model of HoloLens that can be used as a hologram using Maya. The purpose is to teach you how to create the geometry needed for modeling a HoloLens. You could use this model as a base for any other model. You don't have to follow this method exactly. This method only shows you the steps to create the geometry. You are free to use your own artistic talent to decide how the model will be shaped.

Model sheet - base reference images

The first step is to have images of HoloLens from three different directions--front, side, and top views. These images will be imported into Maya within its image plan and will be further used for creating a HoloLens 3D Model. You can download these base images directly from `https://github.com/PacktPublishing/HoloLens-Blueprints` or `https://www.packtpub.com/sites/default/files/downloads/HoloLensBlueprints_ColorImages.pdf`.

- The front view of the HoloLens image looks as follows:

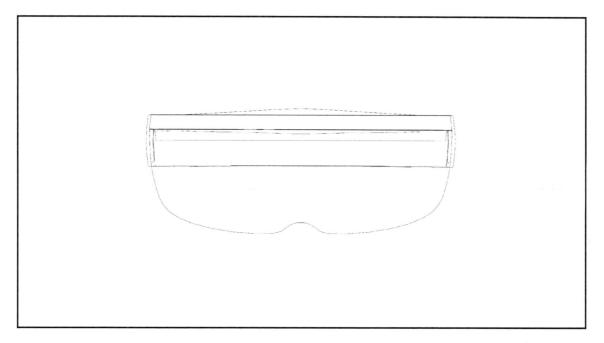

Front view of HoloLens

- The side view of the HoloLens image looks as follows:

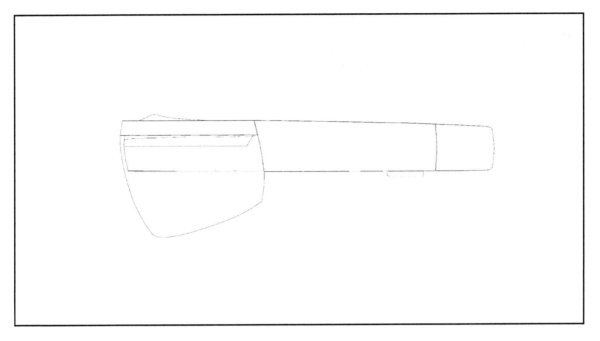

Side view of HoloLens

- The top view of the HoloLens image looks as follows:

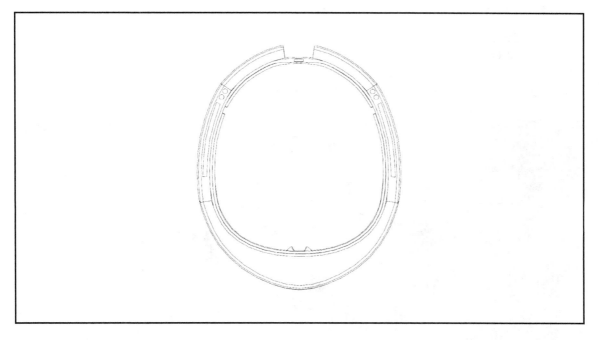

Top view of HoloLens

Scene creation

After your images are ready, let's create a new scene within Maya and import these images. All the 3D Modeling information is saved into a file known as scene. Follow these steps:

1. Click on **File** | **New Scene**.
2. Save the Scene you have just created.
    1. Click on **File** | Save **scene as...**.
    2. Give your scene a name.
    3. Click on **Save**.

Importing model sheet

Within the newly created scene, let's add model sheets by importing base images.

Import side image

Change the view to *Right View* by following these steps:

1. Press Spacebar on your keyboard; that will show shortcut menu options on the screen.
2. Select the **Maya** option, which will bring up more sub-options.
3. From these sub-options, select **Right View.**

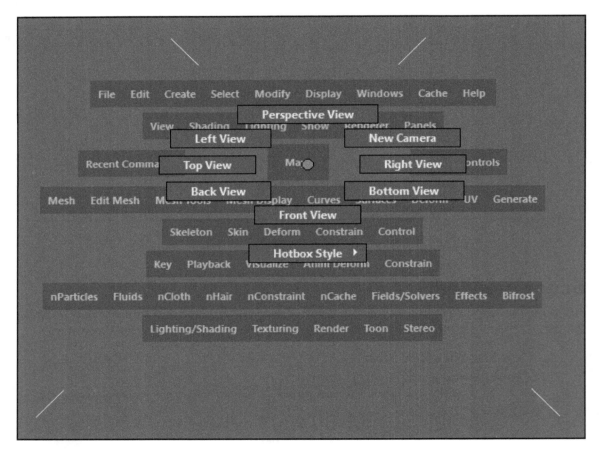

Menu options for changing the views

After the view is changed to **Right View**, the next step is to import an image to this view.

4. Select the **View** menu in the panel's menu bar, and choose **View** | **Image Plane** | **Import Image**:

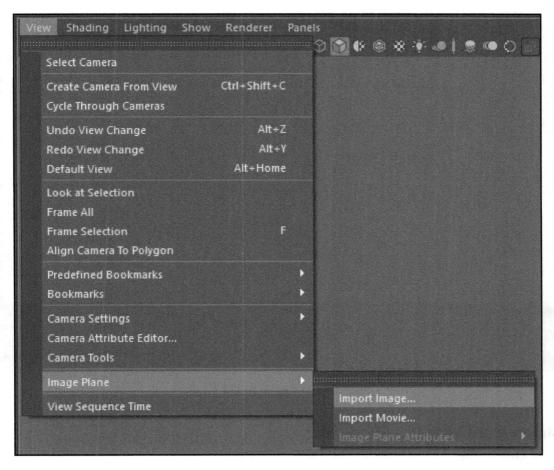

Menu option to import image into the Image Plane

5. Select the `HoloLens_side.png` file and select **Import**:

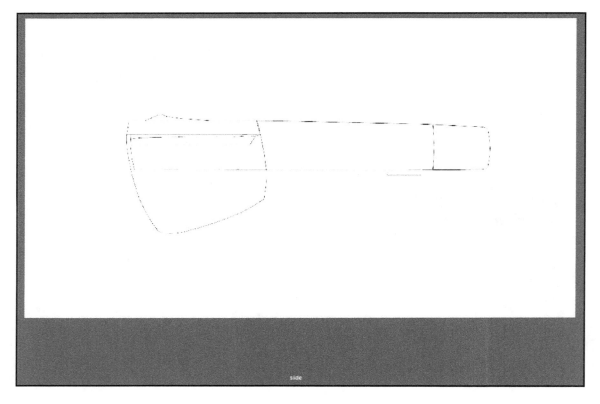

Side View of imported image

Import front image

Now, change the view to *Front View* by following these steps:

1. Press Spacebar on your keyboard, which will bring up shortcut menu options on the screen.
2. Select the **Maya** option, which will bring up more sub-options.
3. From these sub-options, select **Front View.**

After the view is changed to *Front View*, the next step is to import an image to this view.

1. Select the **View** menu in the panel's menu bar, and choose **View | Image Plane | Import Image.**
2. Select the `HoloLens_front.png` file and select **Import**.

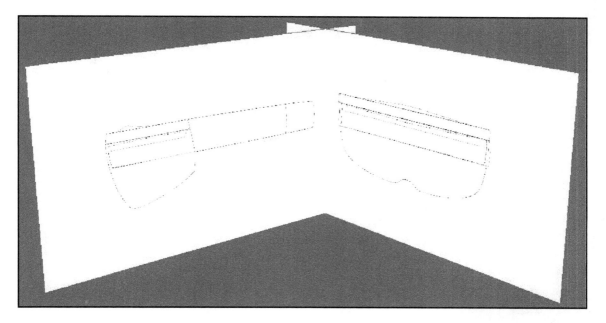

Front View imported into the scene

Import top image

Now, change the view to *Top View*; for this, perform the following steps:

1. Press Spacebar on your keyboard, which will bring up shortcut menu options on the screen.
2. Select the **Maya** option, that will bring up more sub-options.
3. From these sub-options, select **Top View.**

After the view is changed to *Top view*, the next step is to import the image to this view.

1. Select the **View** menu in the panel's menu bar, and choose **View | Image Plane | Import Image.**
2. Select the `HoloLens_top.png` file and select **Import**.

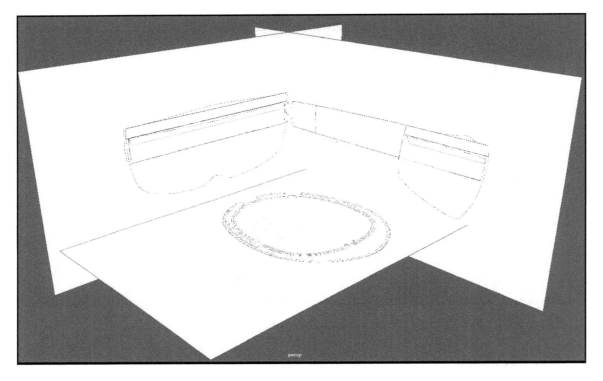

Top View imported into the scene

3D Model construction with cube

After the images have been imported, follow these steps to create a 3D Model of the base frame of the HoloLens:

1. Create a cube by selecting the menu options **Create| Polygon Primitives | Cube**. This will create a standard cube in the center of the grid.

2. Resize this cube, as shown in the following snapshot of cube properties, and reposition it to be placed right in front and at the top of the HoloLens. Refer **Translate X**, **Translate Y**, and **Translate Z** properties for repositioning the cube; refer **Scale X**, **Scale Y**, and **Scale Z** for resizing the cube.

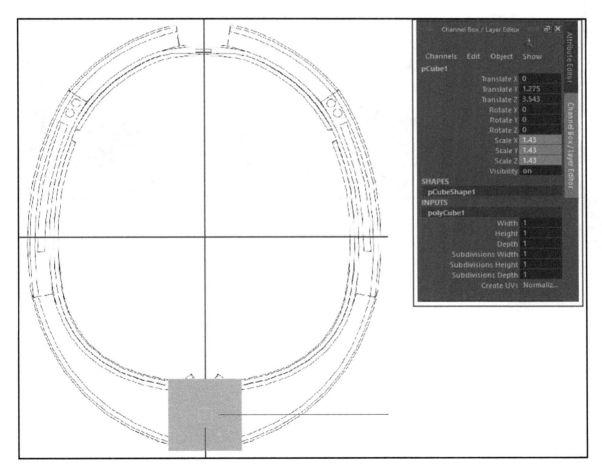

Repositioned cube and cube properties

3. If you change the view to *Perspective View*, by pressing
   Spacebar | **Maya** | **Perspective View**, the scene will resemble the following image:

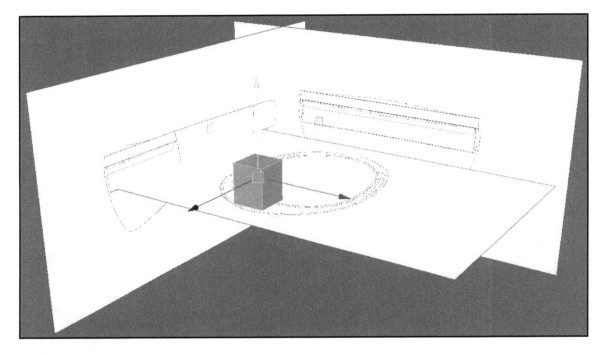

Perspective View of the scene

Edge loop

After the base cube is added, the next step is to split the cube faces by adding an edge loop:

1. Go to the menu option **Mesh Tools | Insert Edge Loop**.

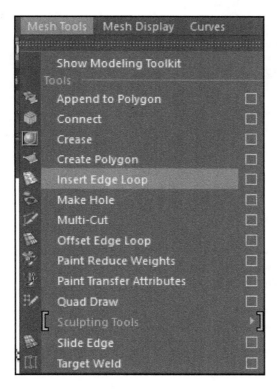

Insert Edge Loop option

2. After inserting the edge loop into the cube, it will look like the following:

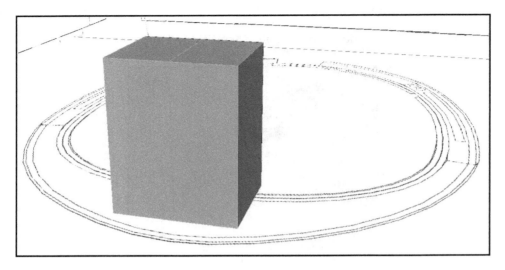

Edge loop in to the cube

3. The next step is to select one side of the face and delete it. For that, select the object in the scene | right-click | short menu options open | select **Face**. This will change the object to component mode.

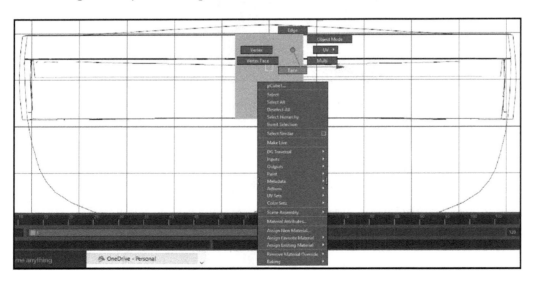

Change to Face mode

4. Now, select one of the faces and press the delete button to delete it.

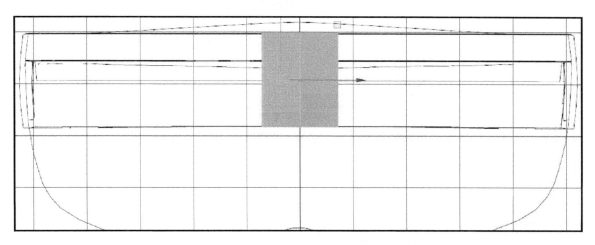

Faces selection

5. After deleting one of the faces, go to face mode by selecting the object in the scene | right-click | short menu options open | select **Face** and select the left side of the face.

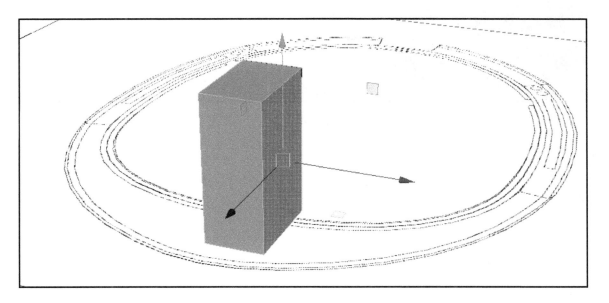

Face selection of the left face of the object

6. The next step is to extrude the face to start building the model. For this, hold *Shift* key | right-click on mouse | select **Extrude Face**.

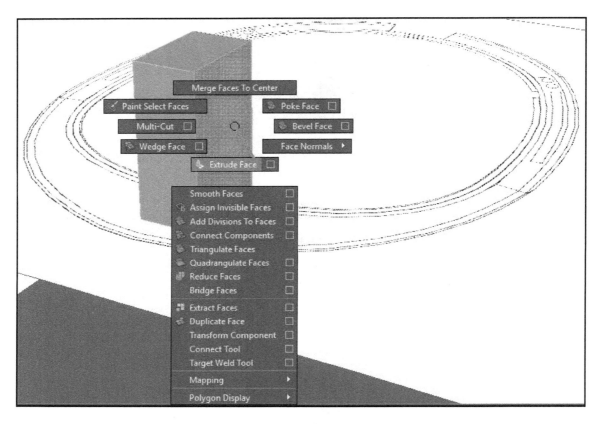

Extrude face option

7. Now, change the **Local Translate Z** option on the properties to 1, and that will create an extruded view.

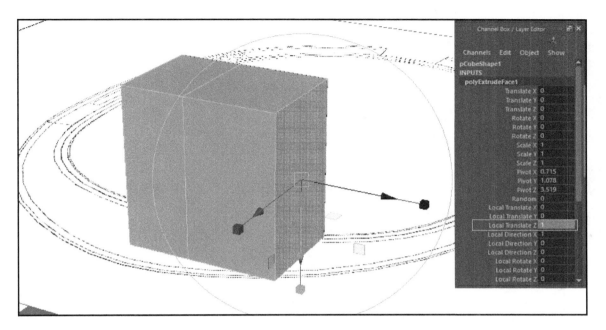

Extruded and translated face view

Vertex adjustment

After an extrusion, you need to change the shape as per the required HoloLens drawing. To achieve that, change to *Top View*. Follow the steps:

1. Right-click on object and select **Vertex**:

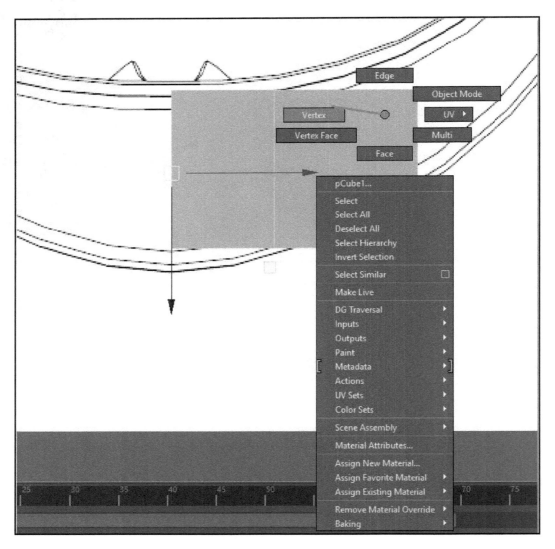

Select Vertex

2. Select **Vertex** and adjust it as per the drawing shape.

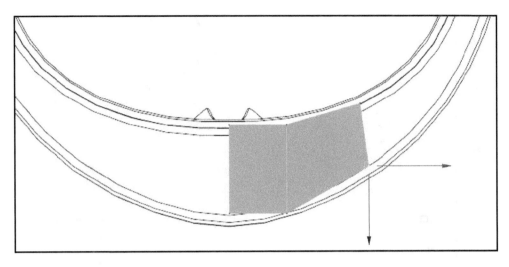

Vertex are adjusted as per the drawing shape

3. In the preceding step, we adjusted the vertex in the top view. Repeat this vertex adjustment exercise for the side and front views also, so that the adjustments match to the drawing in all views.

4. Now, complete this exercise of extrusion and vertex adjustment for one side of the HoloLens shape. After the extrusion exercise, the side view of the 3D Model will look as follows:

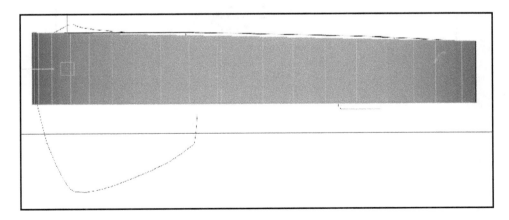

Extrusion and vertex adjustment for one side of the 3D Model--side view

5. The top view of the 3D Model will be as follows:

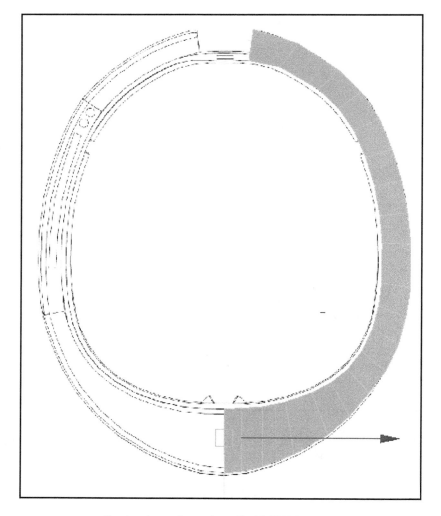

Extrusion and vertex adjustment for one side of the 3D Model--top view

6. The perspective view of the 3D Model will look like the following:

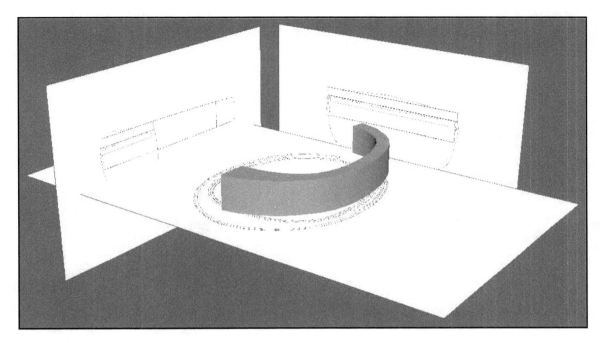

Extrusion and vertex adjustment for one side of the 3D Model--perspective view

Mirroring

Once you have completed one side of the model, you can quickly mirror the object to complete the base model:

1. Go to object mode by pressing the *F8* button.
2. Now we apply the mirror geometry option to duplicate it for the other side. Select menu option **Mesh | Mirror Geometry** and select **-X** as the **Mirror Direction**.

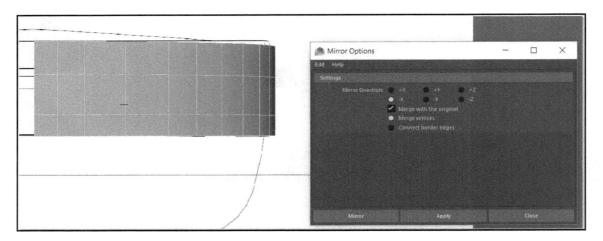

Mirror Options

3. After applying the mirror, you will get the complete base frame ready for the 3D Model. The top view of the 3D Model will look like the following:

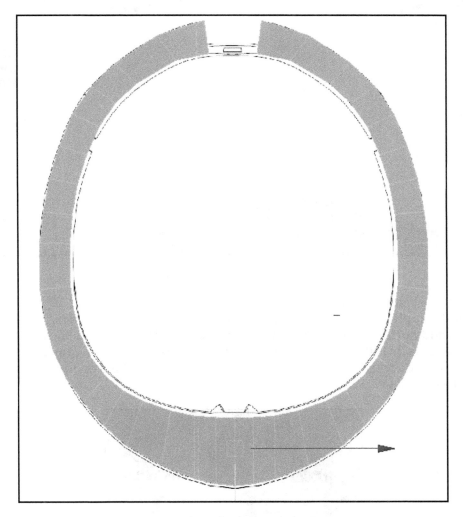

3D Model of the base frame--top view

4. The perspective view of the 3D Model looks as follows:

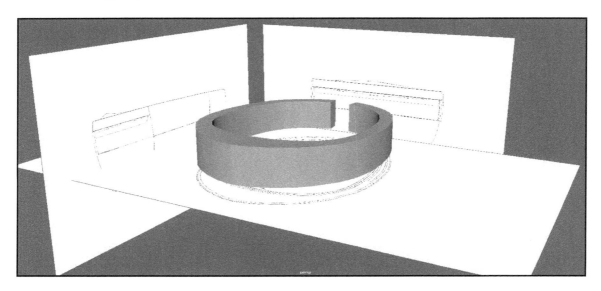

3D Model of the base frame – perspective view

Congratulations! Now, you are ready with your base frame for your first 3D Model. Subsequently, apply this knowledge and start with the base cube, extrude its faces, adjust its vertices, and mirror them to create other components of the required 3D Model. That is, components such as the adjustment band, lenses, speakers, buttons, cameras, and projectors. After developing all these components and placing them on to the scene, you should get a complete 3D Model of the HoloLens device.

Complete 3D Model of the HoloLens device

Texturing workflow

After you have completed basic 3D Modeling, the next step is Texturing. The 3D Model you created in the preceding section looks visually very bland in nature. To make it more realistic in looks, we will apply a process called Texturing. In the texture process, we will take an image and apply it on to the surface of our 3D Model.

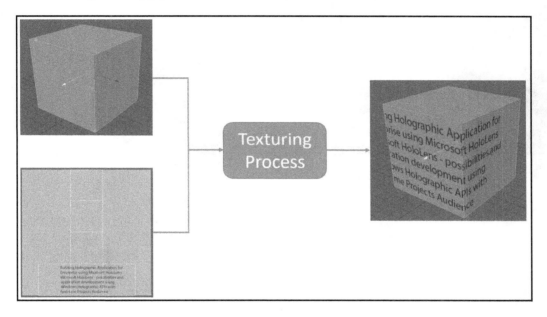

Texturing process

UV Unwrapping of model

Before you go and do Texturing, you must follow a process called UV unwrapping. **UV unwrapping** is unfolding the mesh at the seams and laying out the polygons on a flat page. UV unwrapping can be done manually or automatically, but most of the time we do it manually.

Perform the following steps to achieve UV Unwrapping:

1. Select an object within a scene.
2. Select the menu option **UV | Planar**.
3. Within the **Properties** window, set **Rotate X**, **Rotate Y**, and **Rotate Z** as zero (0) and **Projection Width, Height** as one (1).

4. Select the menu option **UV | UV Editor**.

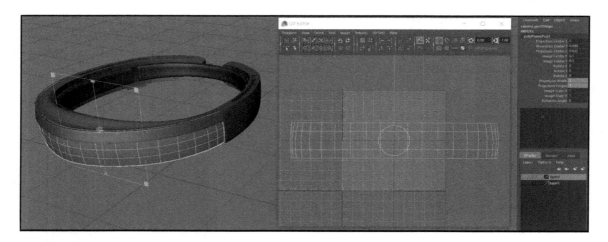

UV Unwrapping of the lens

5. Scale UVs within the range of 0 to 1 for both the x and y axes.
6. Inside **UV Editor**, select the menu option **Polygons | UV Snapshot,** which will open the **UV Snapshot** window.
7. Within the **UV Snapshot** window, select **Path**, where you need to save this image and name the output file.
    1. Define the texture page size, such as **1024 x 1024**.
    2. Select the image format as **PNG.**
    3. Save that file to the location.

Create texture in Photoshop

Once your UV unwrapping is finished, it's time to move onto Texturing. Texturing is an important aspect of the assest pipeline no matter what you're working on, and 3D holograms are no exception. For this, we are going to take a UV snapshot image and use Adobe Photoshop to create a texture image. Within Photoshop, perform the following steps:

1. Create a new file of the same size as a UV snapshot image, which was created in the previous step, that is, of size **1024x1024**.
2. Import the UV snapshot image into the new file.
3. Start creating texture with the help of the UV snapshot.

4. Save this file in the form of PNG; this texture file will be further used while shading in Unity3D.

Textures can be created by editing photographs or using hand-painted textures within Photoshop.

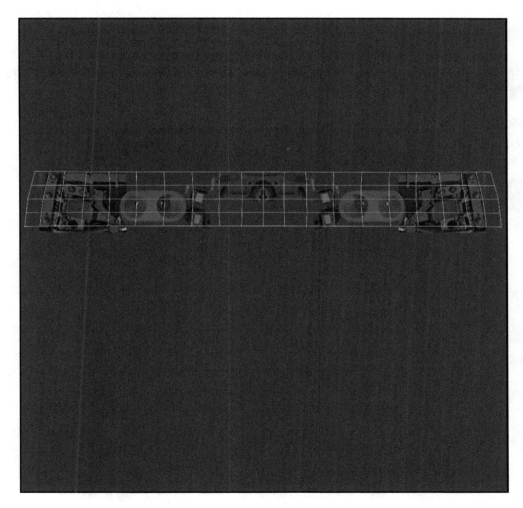

Texture map image output from Photoshop

Export as Filmbox (FBX)

Within Maya, the scene is saved as a .MA file format. The same scene file needs to be exported as an FBX (Filmbox) format. Exporting to FBX has the following advantages:

- It consists of only required content, such as model, material, and so on and discards unwanted nodes
- It increases the reusability of an object, as it maintains Object Hierarchy and granularity, as it's there in the base scene within Maya

While exporting as an FBX file, make sure that you check the **Embed Media** option under **Embed Media**. This includes all the texture nodes in the model file.

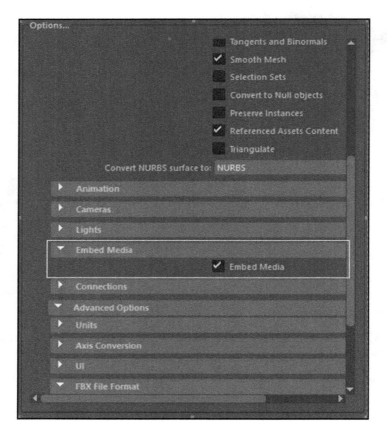

Option before exporting as FBX

Assembling assets in Unity3D

You have already created a 3D Model and exported that in the form of the FBX format. Also, you have a textured image in the form of PNG from Adobe Photoshop. In this section, you will apply this texture file to material, and material will be assigned to surfaces to achieve the final 3D Model. Perform these steps in the Unity3D tool:

1. Create a new 3D project in Unity3D.
2. You can find the `Assets` folder inside the project panel.
3. Create separate folders for each section, such as `Models`, `Materials`, and `Textures`.

Folder hierarchy with Unity3D

4. Import the FBX file, which you generated from Maya, into the `Models` folder by right-clicking on the `Models` folder, and select the context menu option **Import Assets**.
   - The **Import Assets** dialog box will open; select the FBX file and import.

5. Now you will see an FBX file within the `Models` folder. Drag this FBX asset into the scene, and reset **Transform** values.

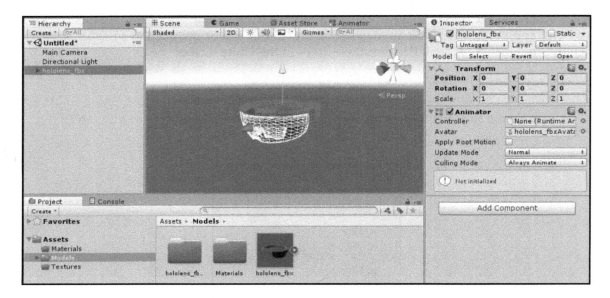

Import of FBX into the Unity3D Scene

6. Now, let's import texture into the scene by right-clicking on the `Textures` folder, and select the context menu **Import New Asset**. Select the PNG file generated from the Adobe Photoshop and **Import**.
7. Now we need to create a new **Shader** and apply texture to it.
    1. Right-click on the `Materials` folder and select **Create | Material** from the context menu.
    2. Select the newly created material and drop it on the object; that will assign the material to the object.

3. Select the object, go to the **Inspector** window and open the **Material** section within it. Now, select texture, which was created in the preceding step, and drag it into the **Albedo** option in the **Material** section of the **Inspector** window.

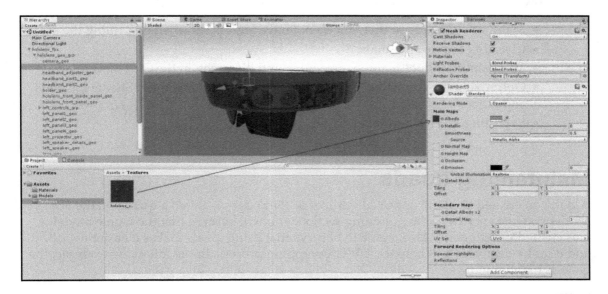

Add texture to the object

8. The asset is ready to be exported and handed over to the development team.
   1. Select the top node of the group in the Hierarchy panel that needs to export.
   2. Select menu option **Assets** | Select **Dependencies.**
   3. Select menu option **Assets** | Export **Package.**
   4. Review the selection with exporting package windows, give a name to the package, and export it.

The development team can import the **Asset** package and start building the app.

In the above section, we talked about Unity3D usage from the perspective of 3D Modeling. In the next chapter, we will talk more in detail about the Unity3D editor from the perspective of the developer and scripting.

Things to consider while using the 3D Asset Pipeline

The following are some points you need to consider while creating 3D Models:

- Maintaining quads so that you do not have any issues when deforming the model
- Checking Surface Normals
- Maintaining proper naming, grouping, and hierarchy in the OUTLINER
- Optimizing the scene file. Delete surface history and unwanted Shaders, Layers, and Nodes
- Exporting in 'FBX' format

Maintaining quads

A polygon created by four connected vertices or four edges is called a **quads**. In the 3D Model you just created in the preceding section, you will observe that the whole model was created using quads. Also, you will observe that using quads makes it easier to edit, and it looks clean and will help the UV Unwrapping process. In the following example, the quad is highlighted:

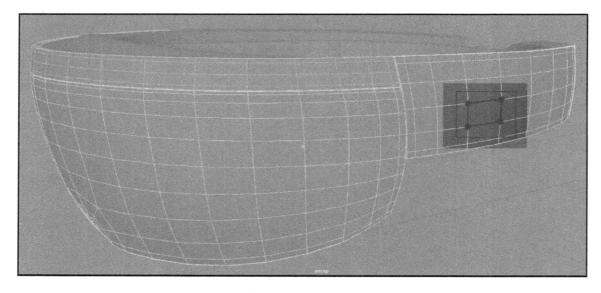

Maintain quads during 3D Modeling

Checking Surface Normals

Each polygon consists of two surfaces, a front surface and a back surface. The front surface is called Surface Normal and is usually represented with an arrow coming out at a 90 degree angle from the surface. It's recommended to have the front surface toward the camera, as usually we apply a single-sided Shader and that gets applied only on the front surface. This also reduces the render time.

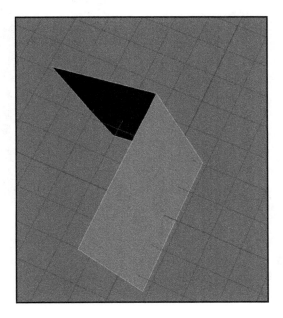

Surface Normal

To reverse the direction of a model's Surface Normals, select the object (or individual faces), press the *Shift* Key | right-click | **Face Normal** | **Reverse Normal**.

Maintaining grouping

A 3D Model consists of a set of different objects which are placed together to form a single node. However, if you put every object in the same hierarchy, then it would be very difficult to maintain, interact, and transform them in that group. That's why we follow a concept called grouping, in which related objects are brought together and put into a single group. Now you can transform, scale, and rotate that group together and easily manipulate them. To put objects into a group, just select all required object, and press *Ctrl + G*; that will put these objects into a new group.

The following is an example, where we have grouped different components of the inner band; this allows freedom to translate, rotate, and scale them together:

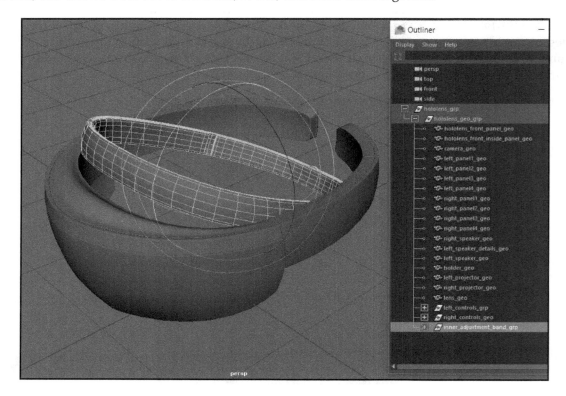

Grouping components together

Optimizing the scene file

While developing, the scene usually consists of objects, shaders, layers, lights, cameras, null objects, and unnecessary surfaces, which may have gotten added into the scene while doing the development.

Use the following best practices to optimize the scene file before sharing it with the development team:

1. If you have textures within your scene, use default display options for polygons by selecting the menu option **Shading** | use **Default Material**.

2. Delete all unwanted nodes, layer, shaders, lights, cameras, and textures.

3. Try to merge the objects together into one (non-animate objects). Maya performs much better with less objects that each have thousands of polygons versus many objects with only a few polygons.

4. Select the model group node in **Export Selection**. Follow the reference image for **Export Selection** settings.

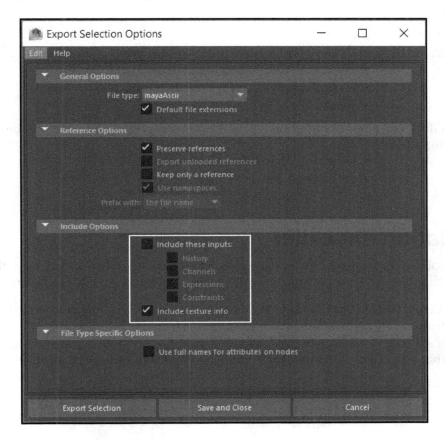

Export selection options

Setting up a development environment

Before going onto the next chapter, start developing a holographic application using assets you developed using the above section; let's first set up the development machine.

Prerequisites

To start the holographic application development, it's recommended to have a machine with at least the following hardware requirements:

- 64-bit Windows 10 Pro, Enterprise, or Education (the Home edition does not support Hyper-V)
- 64-bit CPU
- 8 GB of RAM or more
- In the BIOS, the following features must be supported and enabled:
    - Hardware-assisted virtualization
    - Second-Level Address Translation (SLAT)
    - Hardware-based Data Execution Prevention (DEP)
- GPU (The emulator might work with an unsupported GPU, but will be significantly slower)
    - DirectX 11.0 or later
    - WDDM 1.2 driver or later

Installation checklist

Install the following programs for holographic development. If you already have them, make sure that you have the correct versions installed using HoloLens emulator:

- **Visual Studio 2015 Update 3 (or later)**: Download and install Visual Studio 2015 from `https://developer.microsoft.com/en-us/windows/downloads`. While installing Visual Studio 2015, be sure to opt for a custom installation and ensure that you select the **Universal Windows App Development Tools** option.

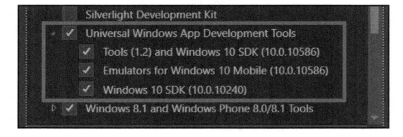

Visual Studio custom installation option

- **HoloLens emulator**: The emulator is a virtual machine which allows you to run an application without the availability of an actual device. Within Visual studio, if you need to test any type of application, such as the Universal Windows Application for tablets or mobile, or Android mobile applications, you will find different types of virtual machines for each type of application. A similar concept, HoloLens emulator is also available, which allows you to test your holographic application without the actual HoloLens device. This makes your development smoother and quicker. Download the HoloLens emulator from `http://go.micro soft.com/fwlink/?LinkID=823018`.

- **Unity3D 5.5**: The latest Unity3D 5.5 release version is 5.5 0f 3 as of November 30, 2016. Download Unity 5.5 or above from `https://store.unity.com/download`.

Using HoloLens emulator

HoloLens emulator will allow you to test holographic applications on your PC without a physical HoloLens device. This emulator uses a Hyper-V virtual machine, which is similar to other emulators, such as windows or android emulators.

To test out the emulator, let's create a sample universal windows holographic application and run it on HoloLens emulator. Follow these steps to create a new holographic application:

1. Open Visual Studio and create a new project by selecting the menu option **File | New | Project**.
2. Within the **New Project** window, select **Holographic DirectX 11 App (Universal Windows)** project type and create P**roject**.

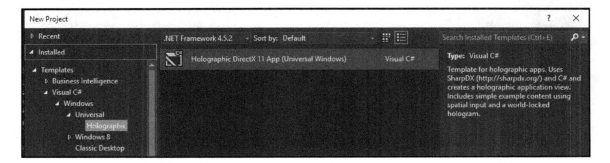

Holographic project option in Visual Studio

3. This will create a default Holographic project with a rotating cube in the default view.

4. Now make sure that **Solution Platform** option is set to **x86**, and **HoloLens Emulator** is selected in the **Emulator** option. After that, build and run the project.

Project type and emulator setting before build

5. After the project is successfully built, it will start the HoloLens emulator, and you will see the following cube window within the emulator:

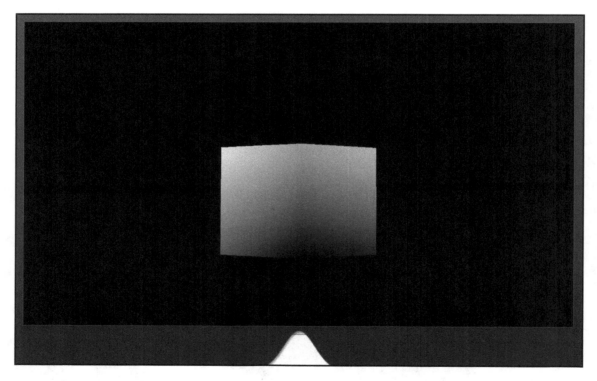

HoloLens emulator running a sample project

Now, let's interact with this holographic application using basic emulator inputs to mimic physical gestures on HoloLens:

Action	Keyboard Input
Walk forward	Use W key on your keyboard
Walk backward	Use A key on your keyboard
Walk left	Use S key on your keyboard
Walk right	Use D key on your keyboard
Look up	Click and drag the mouse or use up arrow keys on your keyboard
Look down	Click and drag the mouse or use down arrow keys on your keyboard
Look left	Click and drag the mouse or use left arrow keys on your keyboard
Look right	Click and drag the mouse or use right arrow keys on your keyboard
Air Tap gesture	Right-click the mouse, and press the Enter key on your keyboard
Bloom gesture	Press the Windows key or F2 key on your keyboard
Hand movement for scrolling	Hold the Alt key, hold the right mouse button, and drag the mouse up / down

Summary

In this chapter, you started the development process of creating your own first holographic application. It began with the envisioning phase, where you learned about requirement gathering process and scenario identification. This was followed by the scenario prioritization and sketching phases, where you learned about prioritization, selection, and sketching scenarios. After that, you learned about 3D Modeling tools, the 3D Model creation process, and finally, setting up a development environment for your first holographic application development.

In the next chapter, you will learn about setting up a development environment, a Unity3D project, importing assets into Unity3D, scripting to implement gestures, building a complete solution, and deploying it on the device.

4
Explore HoloLens as Hologram - Developing Application and Deploying on Device

In the previous chapter, we started our journey of building our very first holographic application using HoloLens. In this chapter, we will continue with the next steps of the application development life cycle from where we left it in the previous chapter. Over the course of these two chapters, we are trying to build training scenarios for HoloLens. We have completed our envisioning phase, where we learned about the requirements and scenario identification for different training scenarios for the HoloLens device. After that, we discussed prioritization of the scenarios to be developed as part of the development phase. We took a step further toward sketching to visualize and revalidate the identified scenarios, which illustrate a kind of vision demonstration of the app. We have also covered the different aspects of 3D holographic model design for HoloLens. Now we have our 3D Assets ready to be used for the development. Finally, we have set up our development environments. All in all, we are all set to take the next steps forward.

This chapter will cover all the steps, from the setting up of the project, to deploying it on the device. A brief overview of the different aspects of this chapter are as follows:

- Setting up the Unity editor for a holographic app
- Understanding the basics of Unity editor
- Setting up the HoloLens toolkit for development
- Interacting with 3D objects using scripting
- Configuring, building, and testing your application
- Various aspects of deployment of your holographic app

We will take a step further toward development with our 3D Assets developed in the previous chapter, and we will give life to the object using scripting in Unity. You will learn how to import different assets and apply scripting to interact with the different components of holographic objects using *Gaze, Gesture*, and *Voice Commands*. Finally, you will learn the different aspects of deployment and run the application on the HoloLens device.

By the end of this chapter, you will have covered the complete end-to-end development process, starting from envisioning to final deployment in the device.

Getting started – creating a new project

To start with, we will create a new project in Unity by following these steps:

1. Run a new instance of Unity3D.
2. Create a new **3D Project** by providing a **Project Name** and all other required details.
3. Click on **Create Project.**

Creating an explore HoloLens project in Unity

This will launch the Unity editor with a newly created project. Before we get into the details of holographic app development, as a developer we must have a basic level of understanding of the Unity editor. Let's have a quick look into some of the important aspects of the Unity environment and some of the Core concepts of Unity that you will be using very often while working with Unity.

The Unity3D basics

In this section, we will quickly have a look at some of the major components of user interfaces for the Unity editor. We will also go through some of the Core concepts of 3D Modelling and Scripting related with Unity3D.

The Core interface

The Core of the Unity3D interface consists of five main panels. They are as follows:

- Scene View
- Game View
- Object Hierarchy
- Project Assets
- Inspector

The following image highlights all the main panels of the Unity editor in a single view:

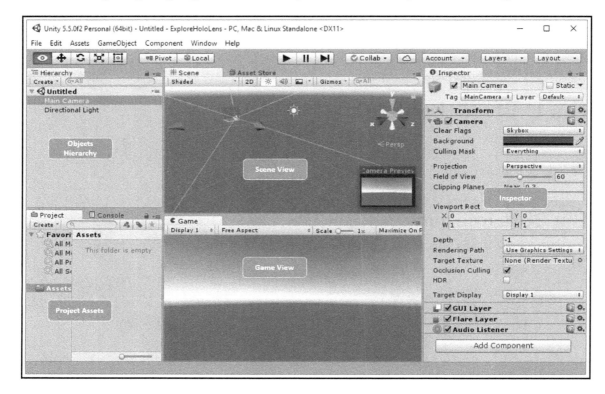

Unity3D Core Interface

Remember, though, all these panels have a default position in the editor; they are floating in nature and you can place and reposition them accordingly as per your need:

Scene View

Scene View is the place where you spend most of the time while working in any Unity project. This is the place where you visually construct and manipulate objects and design the visual aspects of your scene using 2D and 3D Models.

Game View

Game View is used to preview the scene you are constructing in the **Scene View**. This is where you run and test the application within the Unity editor before publishing it for any platforms.

 Game View is just a preview window. Unlike the **Scene View**, you can't select objects in the **Game View**. If you make any changes in the scene when game mode is on, Unity reverts back to those changes as soon as you quite the game mode.

Object Hierarchy

The **Object Hierarchy** panel displays all the objects used in the current Unity scene in a hierarchical order. You can expand and collapse the objects based on the object model depending on their parent and child nodes.

 If you add or remove an object dynamically in runtime using scripts, it will appear on **Object Hierarchy** only when they are active in the current scene.

Project Assets

Project Assets lists all assets that are available and used in the current project including 3D Model, Scripts, Audio, Video and Textures. This is a single place of reference for any assets used for the project.

Inspector

The **Inspector** window or the **Property** window shows the properties of selected objects. This is a context sensitive window and the context depends on which object we have selected. You select objects either from **Scene View**, **Object Hierarchy**, or even from the **Project Assets** window; the **Inspector** window will list the properties.

Select any of the objects, for example, camera and then look at the inspector for different properties of the camera.

The Toolbar

Along with all the five panels in the Unity editor, the buttons on the Toolbar are key to development and manipulating objects. They help you to rotate, scale, and move the object within the scene. The **Play control** helps you to start, stop and pause the game.

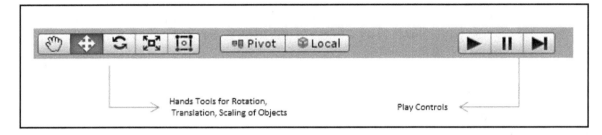

Toolbar controls

The Core concepts

Although you are now familiar with the basics of editor, let us quickly have a look at a few of the Core concepts and terminology that we are going to use over the course of our development. Following is a list of Core concepts of Unity3D:

- Game Object
- Components
- Prefabs
- Tags
- Coordinate system

Game Object

Each individual item in the Unity scene is know as a **Game Object**. You can consider this is as a unit of a Unity scene. A combination of Game Objects used to build up your Unity scene. Be it a 3D Cube, an Audio or a Text; everything within a scene is a **Game Object**.

Components

While a set of Game Objects builds up a Unity scene, components make up the Game Objects, or we can say, each **Game Object** consists of multiple components. There are some predefined components for each **Game Object**, such as **Transform components,** which helps in positioning, rotating or scaling the object.

Apart from all predefined components, there are several other components which can be added and removed as and when required.

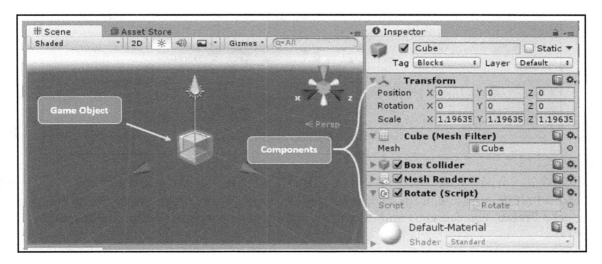

Game Object and its components

 Scripts attached to the Games Objects are also a component of that object.

Prefabs

Prefabs are the key part of development in Unity. **Prefabs** are preconfigured **Game Objects**. When the Game Object is created with all the required configuration and components, then drag and drop it into the Project in `Assets`; your **Prefab** is ready.

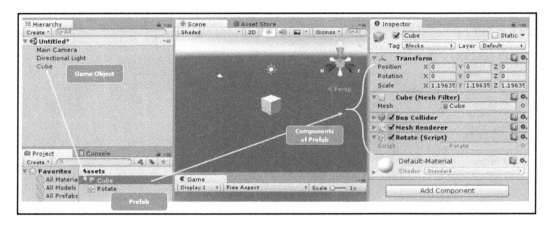

Prefab in Unity

Once a **Prefab** is created, it can be instantiated as a single Game Object on the fly. When you select a **Prefab,** you can view and update the components properties within the **Inspector** panel.

Tags

Tags are used to identify the **Game Objects**. It is like naming a **Game Object**. This is very useful when we try to get a reference of an object from the scripting. We can assign a tag to any object from the **Inspector** window.

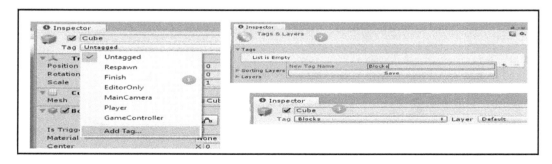

Assign Tags to Game Objects

You can assign the same tag name to multiple Game Objects and you can perform some common operations on all Game Objects using a single tag name.

Coordinate system

Though you can use and design 2D Game Objects, the Core of the Unity coordinate system is a 3D Space. It has the following four coordinate systems, depending on the context that we are using the objects in:

- World
- Screen
- ViewPort
- UI

ViewPort and Screen represent the same area in the scene view, but the way it is represented is different. For example—Screen represents from [0,0] to [Screen.Width, Screen.Height], whereas ViewPort represents from [0,0] to [1,1].

When we add one Game Object to a scene, we use the left-handed coordinate system and the values of the coordinates are as follows:

- Positive X: Right
- Positive Y: Up
- Positive Z: Distance from where user is viewing it

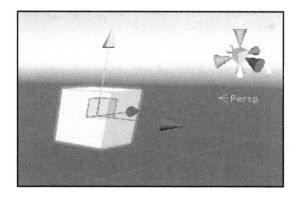

X, Y, Z coordinate for an object

For a left-handed coordinate system, the positive x, y and z axes point in right, up and forward directions, respectively. The Positive rotation is always in a clockwise direction. In the case of the right-handed coordinate system, the positive x and y axes point right and up, however, the negative z axis points forward. In the case of the right-handed coordinate system, Positive rotation is always in a counterclockwise direction.

What we have discussed about Unity3D over here till now will enable you to go ahead with the further development that is required with respect to the book. We will look into the details of each aspect whenever they come in during the development.

In case you want to read more about the details of the editor and different concepts, you can refer to the tutorial at `https://unity3d.com/learn/tut orials`.

Setting up a project for holographic application

We have already created our new project and learned the basics about editors and some of the key concepts that are required to build an application. It's is now time to set up the application for a holographic project-—Explore HoloLens.

Setting up the HoloToolkit - Unity

The HoloToolkit for Unity is a collection of **Unity component**, **Prefab** and **Script** that are used to speed up the development of holographic applications for HoloLens. It supports all types of HoloLens interaction models with the help of separate groups of scripts. Precisely, the HoloToolkit for Unity contains the following areas:

- Input
- Sharing
- Spatial Mapping
- Spatial Sound
- Utilities
- Build

The input module handles all kinds of input related to your holographic apps - starting from *Gaze*, it manages cursor and its related events too. The sharing library allows you to build a distributed application to span across multiple HoloLens. The Spatial Mapping library handles features related with understanding your environment, whereas Spatial Sounds handles audio related features. The `Utility` folder contains several additional extensions and, as the name suggests, it helps to enhance your applications. The **Build** features help you to build, run, and deploy holographic solutions very easily.

We will be using the HoloToolkit for all of our development. We will also write our custom script for specific scenarios as per the requirements.

Get The HoloToolkit - Unity

The following are the steps to be followed for installing and working with HoloToolkit-Unity Package:

1. Download the **HoloToolkit-Unity** Package from GitHub (`https://github.com /Microsoft/HoloToolkit-Unity`).
2. Unzip the downloaded package.
3. From the **Unity** editor | open project and select the folder `HoloToolkit-Unity-master`.
4. Unity will import all the `HoloToolkit` components within the `Assets` folder into your `Project`.

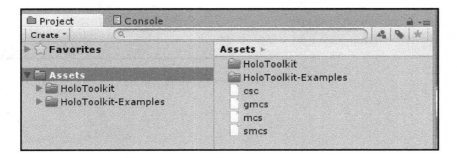

Assets folder within Projects tab

 You may find additional folders here like unit tests for HoloToolkit and so on when you import them. These folders may vary depending on the version of HoloToolkit you are using. We will be only using `HoloToolkit` folder for all of the projects.

Preparing the package

We have two folders `HoloToolkit` and `HoloToolkit-Examples` here. The `HoloToolkit` folder is the one which we are going to use, and the `examples` folder has the reference projects that uses the `HoloToolkit`.

You can go ahead and delete the `HoloToolkit-Examples` and get going on the next step. But, for future reuse, we will create a package only using `HoloToolkit` folder.

In the **Projects** pane, select the `Assets` folder, right-click and select the **Export Package** Option.

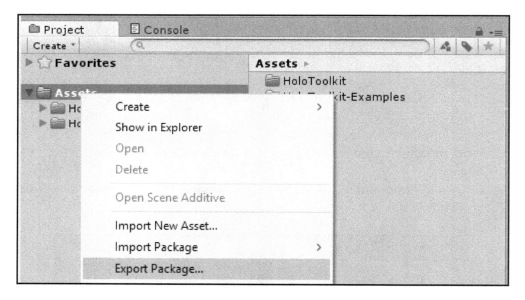

Export Package context menu option

In the next steps, you must select the items to be exported. In this case, we will unselect the `HoloToolkit-Examples` folder and click on the **Export...** button.

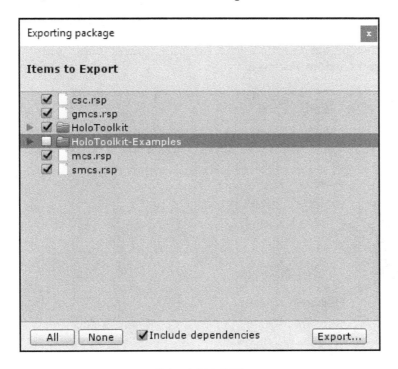

Package choices for export

Finally, save it as **HoloToolkitUnityPackage.unitypackage**.

Created package for HoloToolkit

Your Unity **HoloToolkit** package is ready. Going forward, we will use this package for further development.

Import the package

Now you need to import the **HoloToolkit** package within your **ExploreHoloLens** project. To do that, open **ExploreHoloLens** project and follow the steps below:

1. Right-click on the `Assets` folder, from the context menu **Import Package** | **Custom Package.**

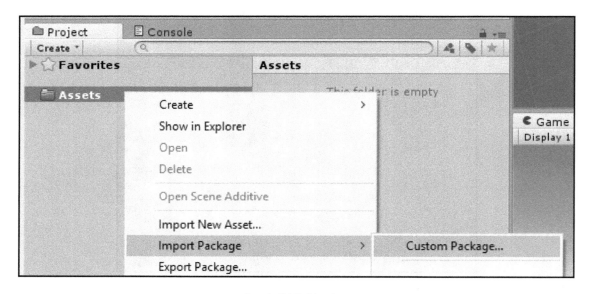

Importing HoloToolkit package

2. The **Import Package** window will list all the script components that are getting imported. You do not need to change anything at this point in time, just click on **Import.**

Importing Package in Unity

3. Unity will take a couple of minutes to load everything and you are all done with the package import.

 The number of folders and files may vary depending on the version you are using. During the course of this book exercise, we will use the same package we just created and would recommend using the same. You can download the package as part of the solution package of this chapter.

Applying project settings

Once the package is installed, you should now have a **HoloToolkit** as a menu item.
Navigate to **HoloToolkit** | configure and select **Apply HoloLens Scene Settings** and **Apply HoloLens Project Settings**.

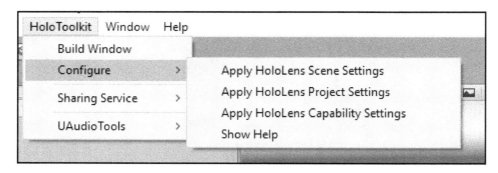

HoloLens menu options

Unity will ask for a reload and let us do that.

Behind the scenes

By applying these two settings, we ensure all the **Settings for Holographic App** setup is completed. Where it enables the platforms and other holographic app specific settings along with enabling the MR capabilities in **Project Settings**.

In case you have to do it manually rather than by selecting **Configure** options, navigate to **Build & Settings** options from **File** menu and set the properties as selected below:

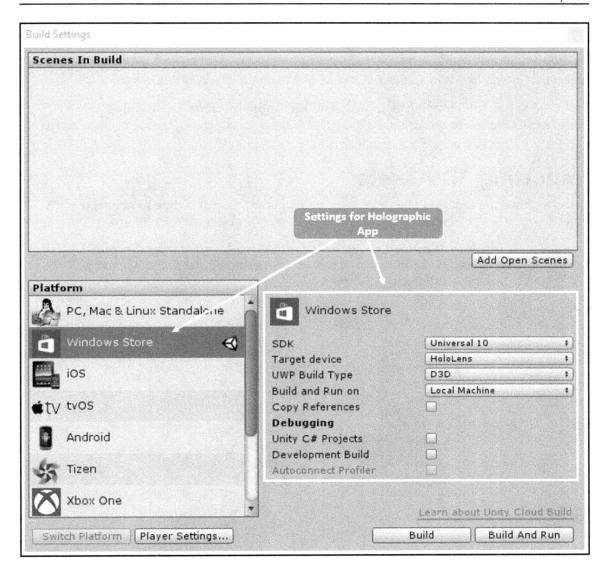

Build and Run Configuration for HoloLens

Updating the initial scene

In the default screen, you will only have the **Main Camera** and **Directional Lights** game objects. Remove the **Main Camera** Game Object from the **Object Hierarchy**.

From the `Asset` folder, navigate to **HoloToolkit | Input | Prefabs** and add the **HoloLensCamera** into the scene.

Importing 3D Assets

Once we have the required objects present in the game scene, it is time to import the 3D objects we created in the previous chapters--Yes, the 3D objects of a HoloLens.

 You can download the FBX file for the HoloLens from for this project from a shared <**location**>. Note, this FBX is nothing but a prefab we have created and discussed earlier.

To import it into our project, right-click on `Assets` | **Import New Assets** and select the FBX file from the download location.

Import HoloLens FBX File

Once the import is done, drag and drop the HoloLens object to Object Hierarchy.

- Scale up the object to 30 for the scale position to get a room size view
- Change the Rotation of the Y axis to -120 degree for a better view from the side

Finally, this is how your initial setup should look in the Unity Scene View and **Game View**.

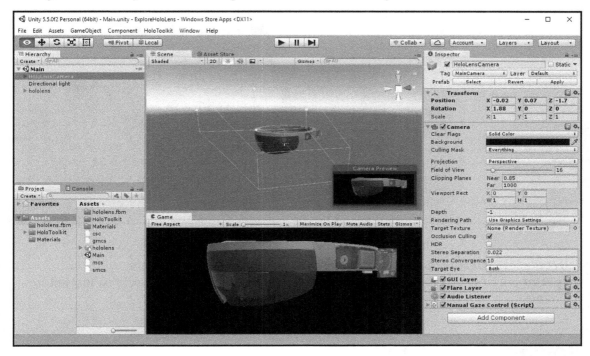

Initial View of the scene

Exploring the 3D Object

Once you import the 3D Model of the HoloLens, you can explore the hierarchy inside the **Object Hierarchy** browser. It has consisted of several smaller Game Objects, as you have seen in the previous chapter. For development purposes, wherever we refer to any of the objects, you must refer them from the **Hierarchy** browser.

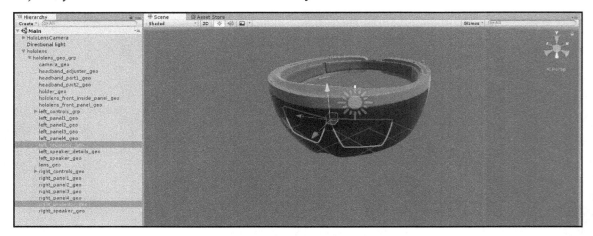

HoloLens 3D object in Object Hierarchy window

Select the **left_speaker_details_geo** object in object explorer, and from the **Inspector** window deactivate the object as of now by unchecking the checkbox just before the name.

Saving the scene

After the basic setup and object insertion is done, let us go to **File | Save Scene** by providing a scene name, let's say **Main.**

We are all set with our project setup and the app is ready to build and run for our holographic application.

Build, run and test the initial setup

When our basic setup for the project is done, we can just build and run the application to test if everything is working fine. In the previous chapter, we already talked about setting up the development environment and tested the emulator with a simple holographic application. Here, we have already set project settings in the **Apply Project Settings** section of this chapter. To cross verify it, from the main menu select **File | Build Settings** and verify all the build settings or update the values as below:

- **Platform | Windows Store**
- **SDK | Universal 10**
- **Target Device | HoloLens**
- **UWP Build Type | D3D**
- **Build and Run on | Local Machine**

1. For local development reference, you can also check the **Unity C# Projects** and **Development Build** checkbox.
2. Click on **Add Open Scene** button and select the checkbox associated with the **Main** scene from the **Scenes in Build** section.
3. Finally, from the **Build Settings** window, select the **Player Settings | Others Settings |** and select the check box for **Virtual Reality Supported.**
4. Now, in the **Build Settings** window, click **Build**.
    1. Create a **New Folder** named `ExploreHoloLensApp`.
    2. Single-click the `ExploreHoloLensApp` folder.
    3. Press **Select Folder**.

From here, Unity will take it forward and perform the project builds and will create the solution for Visual Studio. Remember, all the holographic app is when Unity is done, a **File Explorer** window will appear.

1. Navigate to the `ExploreHoloLensApp` folder and open the **Visual Studio Solution** (the `ExploreHoloLensApp\ExploreHoloLens.sln` file) created by Unity.
2. In Visual Studio, change the build target from **Debug** to **Release** and the platform target from **ARM** to **X86**.
3. Select **HoloLens Emulator** from the list of run options.
4. From the main menu, click **Debug | Start Without debugging** or just press *Ctrl + F5*.

Wait till the HoloLens emulator starts and your application launches. Once the app is launched, you should be able to see the 3D Model of your hologram, as shown in the following image:

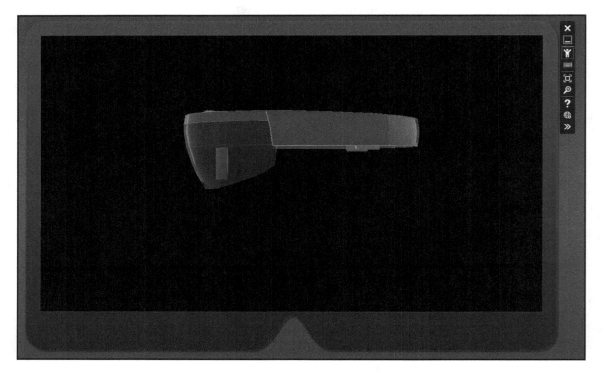

HoloLens hologram within emulator

Giving life to object – scripting

Now let's get into the scripting part. We will be using C# language for the script. In this section, we will discuss the following interactions with holograms:

- **Gaze**: Highlighting and selecting objects
- **Air Tap:** Applying an action on the selected object
- **Voice Command:** Invoking an action through Voice Command
- **Speech:** Response back from the hologram

Highlighting the objects - enabling Gaze

In the previous chapter, we discussed different forms of input and interaction models for HoloLens. Gaze is one of the main forms of input on HoloLens, which indicates where the user is currently looking. With Gaze we will highlight the following two components of our holographic object:

- Lenses
- Speaker

The Gaze manager

Once the user gazes at an object, a cursor will indicate the gazed object. To enable the gaze and have a cursor in your scene, we will use components from the HoloToolkit.

Following are the steps:

1. Add an Empty Game Object in the **Object Hierarchy** and rename it **Root.**
2. Go to the `Assets` folder, navigate to **HoloToolkit I Input I Scripts.**
3. Drag and drop the **InputManager** script to **Root** object.
4. Go to the `Assets` folder again, navigate to **HoloToolkit I Input I Scripts I Gaze.**
5. Drag and drop the **GazeManager** and **GazeStabilizer** script to the **Root** object.

The `GazeManager` class manages everything related with gazing at an object based on `RaycastHit`. The `GazeStabilizer` class helps in stabilizing the Gaze. It samples the form several `RaycastHit` positions and helps in stabilizing your gaze for precision targeting. The `InputManager` class manages the source of Inputs.

The Cursor

The Cursor is used to provide an indication to the object which is currently being gazed at. You can use any object for an indicator, such as arrow, circle, and so on. In this case, we will use the reference cursor from **HoloToolkit**. To add a Cursor into the scene, follow the steps below:

1. Navigate to `Assets` I **HoloToolkit I Input I Prefabs I Cursor.**
2. Drag and drop the **Cursor** prefabs in the **Object Hierarchy.**

The cursor maintain states depends on when it is On Hologram, Off Hologram or even it is interacting with Hologram; very similar to how a mouse cursor behaves when we do a mouse hover or click. The script attached to the Cursor Game Object (*ObjectCursor*) takes care of everything, along with interacting with `GazeManager`.

Adding the Box Collider

Once the `GazeManager` class and Cursor is in place, the next step is to identify the gaze able objects. As part of the requirement for this project, we discussed highlighting the Lenses and Speaker. To gaze at an object, the very first thing you need to do is add a Box Collider to the object. The Box Collider is one of the basic collision primitives in Unity. The ray can happen on the basis of the Collision on the object that you are looking at.

Our objective is to gaze on the Lenses and left-side speaker. So, our first job would be identifying them from Object Hierarchy and then adding Box Collider components.

1. Expand the HoloLens Game Object from **Object Hierarchy**.
2. Select **lens_geo.**
3. Navigate to **Inspector Window | Add Components | Add Box Collider.**
4. Similarly, find the `left_speaker_geo` object, Add the **Box Collider** from *Add Components* as well.
5. Adjust the size of **Box Collider** based on the area where you want gaze cursor to appear.

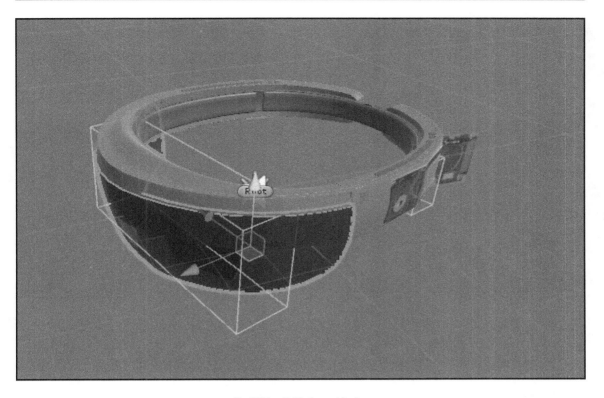

Box Collider added for Lens and Speaker

 For a basic Unity object like Cube, Sphere are having Box Collider components added by default. So we don't need to add Box Collider explicitly to gaze to work.

See Gaze in action

Save the scene, build and run the solution in emulator once again. Now you should be able to see the Cursor as soon as you Gaze on either Lens or in to the left speaker, as shown in the image below:

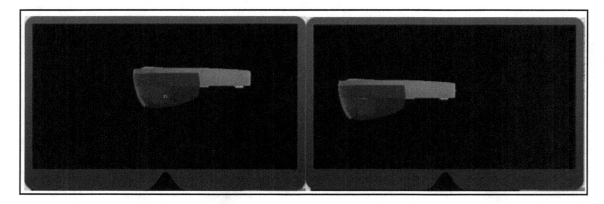

Gaze and Cursor on Lens and Speaker

If you look around the device apart from where we intended to have the cursor by placing the Box Collider, you should see only a basic indication. However, the cursor will only be visible to those intended objects. That is how the different state of cursor works when it is on a gazeable object and when it is not.

Behind the scene

The entire things is managed by `GazeManager` and `Cursor`. On each frame update, the `Update()` method for `GazeManager` is invoked calling the `UpdateGazeInfo()` method to update the current gaze information. The `Update()` method also invokes the `RaycastPhysics()` to determine the gaze position. This method performs a Unity physics Raycast to determine which objects with a collider is currently being gazed at. The `RaycastUnityUI()` method performs a Unity UI Raycast and compares it with the latest 3D Raycast, and finally overwrites the hit object info if the UI gets focus after Gaze. On the other hand, the `Cursor` class manages all the state transformation of the cursor based on the gazed objects.

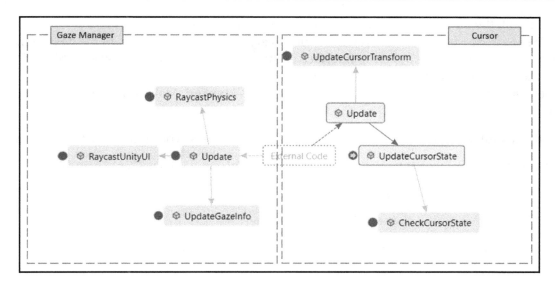

Gaze Manager and Cursor in Code Map

With this we are done with the first part of our journey. The next task will be applying the gesture, where we will do an Air Tap to expand the object and see the inner details of it.

Adding actions to objects - applying gesture

Once gaze is in place and we can see the cursor on the intended objects, we need to invoke the gesture action. As per the requirement, when we do an Air Tap, the object needs to be expanded to view the inner details. So, let's make the Air Tap in action.

Adding Air Tap on Lenses

To add an Air Tap gesture, we need to write a custom script for it. But, before that, Go to the `Assets` folder, navigate to **HoloToolkit | Input | Scripts | InputSource**, and Attach the `GestureInput.cs` with **Root** Object in the **Object Explorer**. Then perform the following action for writing custom scripts:

1. Create a folder called `Scripts` inside the `Assets` folder.
2. Add a new script by navigating from the context menu **Create | C# Scripts**, naming it `LensGestureHandler`.
3. Open the script file in Visual Studio.

By default, the `LensGestureHandler` class is inherited from `MonoBehaviour`. `MonoBehaviour` is the base class from which every Unity script derives. In the next step, Implement `InputClickHandler` interface in the `LensGestureHandler` class. This interface implement the methods `OnInputClicked()` that invoke on click input. So, whenever you do an Air Tap Gesture, this method is invoked. Make sure you include using `HoloToolkit.Unity.InputModule` namespace.

So, the default skeleton for the `LenseGestureHandler` class would look like the following:

```
public class LensGestureHandler : MonoBehaviour, IInputClickHandler
{
    public void OnInputClicked(InputEventData eventData)
    {
        throw new NotImplementedException();
    }
    // Use this for initialization
    void Start()
    {
    }
    // Update is called once per frame
    void Update()
    {
    }
}
```

What we want to achieve by the Air Tap event is to expand the inner lenses. To do that we need to first identify the specific objects from **Object Hierarchy**.

1. In the **Object Hierarchy**, find out these three **holder_geo, left_projector_geo, right_projector_geo** objects.
2. Select them individually and tag them in **Inspector** window **ImuHolder, LeftProjector** and **RightProjector** respectively.

 In **Inspector** windows, from the **Tag Drop Down, Use Add Tag** option to add new tags. Once new tags are added, map them with respective controls for tagging.

The purpose of having this tagging is to identify the Game Object easily using script. The following images show when all these three objects are selected together. Our object is to pull these objects down together when Air Tapped:

Selected objects on 3D Model

Add a new method `TranslateLenseObjects` in the `LensGestureHandler` class, which transforms all three objects and adjusts their y-position. `GameObject.FindWithTag()` finds the specific object by the tag name and then we are applying translation upward or downward based on the value of Y passed as parameter.

```
private void TranslateLenseObjects(float y)
    {
        var lftprojector = GameObject.FindWithTag("LeftProjector");
        var rgtprojector = GameObject.FindWithTag("RightProjector");
        var imuhlder = GameObject.FindWithTag("ImuHolder");

        lftprojector.transform.Translate(0.0f, y * Time.deltaTime,
        0.0f);
        rgtprojector.transform.Translate(0.0f, y* Time.deltaTime,
        0.0f);
        imuhlder.transform.Translate(0.0f, y * Time.deltaTime, 0.0f);
    }
```

Update `OnInputClicked` method with following code base. This code is self-explanatory. `GazeManager.Instance.HitInfo` returns the `HitInfo` property with details of the gazed object. When the Raycast object has some value, which indicates it has gazed and we are calling the `TranslateLenseObjects()` method with a negative value of Y (-0.5f) and making the tapped is true. In cases when not tapped, we are transforming the object back again to the same position by passing a positive value of Y (0.5f) to the `TranslateLenseObjects()` method.

```
RaycastHit hit;
bool isTapped = false;
public void OnInputClicked(InputEventData eventData)
    {
        hit = GazeManager.Instance.HitInfo;
        if (hit.transform.gameObject != null)
        {
            isTapped = !isTapped;
            if (isTapped)
            {
                TranslateLenseObjects(-5.0f);
            }
            else
            {
                TranslateLenseObjects(5.0f);
            }
        }
    }
}
```

Finally, attach the `LensGestureHandler` script with *lens_geo object*. That's all that is required for Air Tap on Lenses and their transformation on Air Tap.

The `IFocusable` interface works when the focus of gaze is in or out. You can implement this interface with any Gaze Handler which implements `OnFocusEnter()` and `OnFocusExit()` methods. You can use this method for any additional operation or animation and so on.

```
public void OnFocusEnter()
{          // Do something when Gaze In      }
public void OnFocusExit()
{          // Do something when Gaze Out      }
```

Air Tap on Lenses - see it in action

Save the scene, Build and Run the solution in emulator once again. You should be able to see the cursor as soon as you Gaze on Lenses, then do an Air Tap on Lenses when it is gazed and tap it back afterwards.

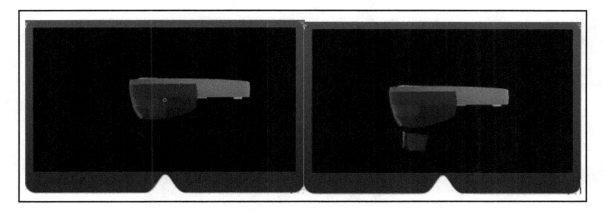

Air Tap on Lenses

Adding Air Tap on speaker

In this project, we will be using the left-side speaker for applying Air Tap on speaker. However, you can apply the same for the right-side speaker as well.

Similar to Lenses, we have two objects here which we need to identify from the object explorer.

- Navigate to **Left_speaker_geo** and **left_speaker_details_geo** in **Object Hierarchy** window
- Tag them as **leftspeaker** and **speakerDetails** respectively

By default, when you are just viewing the Holograms, we will be hiding the speaker details section. This section only becomes visible when we do the Air Tap, and goes back again when we Air Tap again:

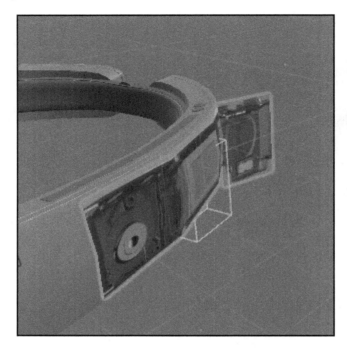

Speaker with Box Collider

- Add a new script inside the `Scripts` folder, and name it `ShowHideBehaviour`. This script will handle the Show and Hide behaviour of the `speakerDetails` Game Object.

Use the following script inside the `ShowHideBehaviour.cs` file. This script we can use for any other object to show or hide.

```
public class ShowHideBehaviour : MonoBehaviour
{
    public GameObject showHideObject;
    public bool showhide = false;
    private void Start()
    {
        try
        {
            MeshRenderer render =
                showHideObject.GetComponentInChildren<MeshRenderer>();
```

```
            if (render != null)
            {
                render.enabled = showhide;
            }
        }
        catch (System.Exception)
        {
        }
    }
}
```

The script finds the MeshRenderer component from the gameObject and enables or disables it based on the showhide property. In this script, showhide is property exposed as public, so that you can provide the reference of the object from the Unity scene itself.

Attach ShowHideBehaviour.cs as components in speakerDetails tagged object. Then drag and drop the object in the showhide property section. This just takes the reference for the current speaker details objects and will hide the object in the first instance.

Attach show-hide script to the object

By default, it is unchecked, showhide is set to false and it will be hidden from view. At this point in time, you must check the **left_speaker_details_geo** on, as we are now handling visibility using code.

Now, during the Air Tapped event handler, we can handle the render object to enable visibility.

1. Add a new script by navigating from the context menu **Create | C# Scripts**, and name it SpeakerGestureHandler.
2. Open the script file in **Visual Studio**.

3. Similar to `SpeakerGestureHandler`, by default, the `SpeakerGestureHandler` class will be inherited from the `MonoBehaviour`. In the next step, implement the `InputClickHandler` interface in the `SpeakerGestureHandler` class. This interface implement the methods `OnInputClicked()` that invoke on click input. So, whenever you do an Air Tap gesture, this method is invoked.

```
RaycastHit hit;
bool isTapped = false;
public void OnInputClicked(InputEventData eventData)
  {
      hit = GazeManager.Instance.HitInfo;
      if (hit.transform.gameObject != null)
      {
          isTapped = !isTapped;
          var lftSpeaker = GameObject.FindWithTag("LeftSpeaker');
          var lftSpeakerDetails =
              GameObject.FindWithTag("speakerDetails");
          MeshRenderer render =
              lftSpeakerDetails.GetComponentInChildren
                  <MeshRenderer>();

          if (isTapped)
          {
              lftSpeaker.transform.Translate(0.0f, -1.0f *
                      Time.deltaTime, 0.0f);
              render.enabled = true;
          }
          else
          {
              lftSpeaker.transform.Translate(0.0f, 1.0f *
                      Time.deltaTime, 0.0f);
              render.enabled = false;
          }
      }
  }
}
```

When it is gazed, we find the Game Object for both `LeftSpeaker` and `speakerDetails` by the tag name. For the `LeftSpeaker` object, we are applying transformation based on tapped or not tapped, which worked like what we did for lenses. In the case of speaker details object, we have also taken the reference of `MeshRenderer` to make it's visibility true and false based on the Air Tap. Attach the `SpeakerGestureHandler` class with `leftspeaker` Game Object.

Air Tap in speaker – see it in action

Air Tap action for speaker is also done. Save the scene, build and run the solution in emulator once again. When you can see the cursor on the speaker, perform Air Tap.

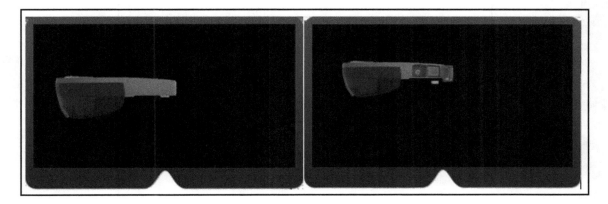

Default View and Air Tapped View

Behind the scenes

Gesture is linked with Gaze. When the user Air Taps on an object, it will return the `HitInfo` property only of the object Gazed on at that point of time. The `HitInfo` property is checked and updated by `GazeManager`. Air Tap on an object if `HitInfo` does not refer to any Game Object, the Air Tap will be skipped and no action will be taken. Input clicked event is driven by the `GestureInput` and `InputManager` class. `OnInputClicked()` method in `SpeakerGestureHandler` class finally handles the operation for a click event.

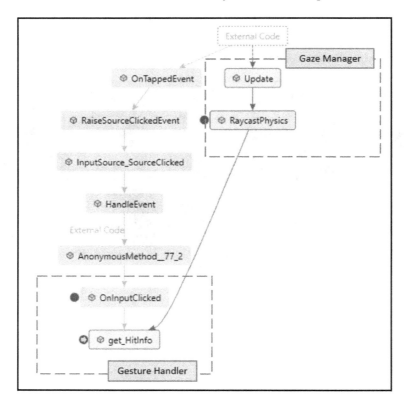

Gesture handler in code map

Interaction using Voice Command

Voice is one of the main inputs for your holographic application. While building any holographic application, we should consider adding the Voice Command for the same. In this application, we will try several Voice Commands to interact with.

Enabling the Microphone capabilities

First things first--Voice Commands are accepted by HoloLens using the Microphones. Providing access to specific capabilities, we let our application allow specific hardware components. Enabling access to the Microphone allows the microphone to listen and capture the voice from the user and pass it on to the application.

From the Unity main menu, **HoloToolkit** | **Configure** | **Apply HoloLens Capability Settings.**

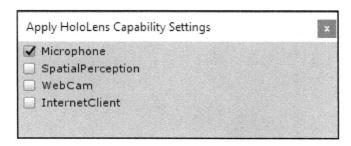

Applying HoloLens Capability Settings

Check the **Microphone** option to enable Microphone capabilities. And finally, save the scene.

 You can also enable this setting by navigating from **Edit** | **Project Settings** | **Players and Under Settings for Windows Store** | **Publishing Settings** | **Capabilities**, you will find option **Microphone**. When **Applying HoloLens Capabilities Settings** from HoloLens Toolkit, it enables the same option only (Enable Microphone Capabilities).

The Keyword Manager

Keyword Manager is the key for Voice Command when we are using HoloToolkit. Keyword Manager allows us to specify keywords and register the methods to be executed for specific keywords in the Unity Inspector. We don't need to write explicit code for this.

From `Assets`|**HoloToolkit** | **Input** | **Scripts** | **Voice**, include the Keyword Manager as a component for the Root object.

The Keyword Manager also includes a setting to either automatically start the keywords recognizer or allow your code to start it. In our case, we will start it automatically by setting up the `Recognizer Start` option as `Auto Start` in the **Inspector** window.

Keyword Manager uses a set of keywords to match with and performs actions corresponding to each keyword.

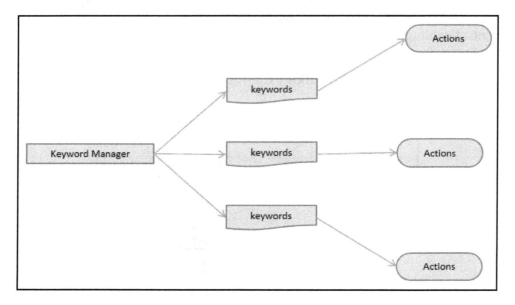

Keyword Manager , Keywords and their actions

Defining Keywords

Once the Keyword Manager is included in the Game Object, our next job will be defining the keywords--which are nothing but Voice Commands that the user is going to say. As part of the requirement discussion, we discussed using Speech Command:

- **Display Speaker Details**: It's the same action for Speaker we did using Air Tap. We will be using two Keywords for this case:
  - Show Speaker
  - Hide Speaker

- **Rotate the HoloLens Object**: Voice Command to Start and Stop Rotate the entire object in front of you, so that you can take a 360-degree view from a single place. Following are the two keywords for these specific scenarios:
  - Start Rotate
  - Stop Rotate

Implementing actions for Keywords

Once we have our keyword defined, we need to implement the action for respective keywords. For show hide speaker, our Action will be the same as we did by Air Tapping in Speaker. However, in this case, we need two different action methods `ShowSpeaker()` and `HideSpeaker()` that will get called respectively when we invoke the Show Speaker and Hide Speaker Voice Commands respectively.

1. Add a new script called `VoiceCommandHanlder.cs` inside `Assets| Scripts`.
2. Replace the default code with following lines of code:

```
public class VoiceCommandHandler : MonoBehaviour
{
    private bool isShowing = false;

    public void ShowSpeaker()
    {
        if (!isShowing)
        {
            var lftSpeakerDetails =
                GameObject.FindWithTag("speakerDetails");
            MeshRenderer render =
                lftSpeakerDetails.GetComponentInChildren
                <MeshRenderer>();
            gameObject.transform.Translate(0.0f, -1.0f *
                Time.deltaTime, 0.0f);
            render.enabled = true;
            isShowing = true;
        }
    }

    public void HideSpeaker()
    {
        if (isShowing)
        {
            gameObject.transform.Translate(0.0f, 1.0f *
                Time.deltaTime, 0.0f);
            var lftSpeakerDetails =
                GameObject.FindWithTag("speakerDetails");
            MeshRenderer render =
                lftSpeakerDetails.GetComponentInChildren
                    <MeshRenderer>();
            render.enabled = false;
            isShowing = false;
        }
    }
}
```

This code block is doing the same things as the Air Tap on speaker. First, get the `speakerDetails` Game Object reference by the tag name, get the reference of the `MeshRenderer` of child components and show and hide it for the respective commands. Also, Transform the Game Object in *y* axis based on the command patterns.

Finally, attach the `VoiceCommandHandler.cs` with `Left_Speaker_geo` Game Object.

For Rotation of the entire object, we will write a custom script. When we call the method `StartRotate()`, the object will start rotating . `StopRotate()` will stop the rotation.

1. Add a new script called `RotateVoiceCommandHandler.cs` inside `Assets | Scripts`.
2. Replace the code with following lines of code.

```
public class RotateVoiceCommandHandler: MonoBehaviour
{
    private bool isRotating = false;

     // Method for Start Rotate Voice Command
    public void StartRotate()
    {
        isRotating = true;
    }

     // Method for Stop Rotate Voice Command
    public void StopRoate()
    {
        isRotating = false;
    }

// This method gets call for each frame.
// Depends on Rotation Status (isRotating) it object will be
        transformed

    void Update()
    {
        if (isRotating)
        {
            transform.Rotate(Vector3.up * Time.deltaTime,
Space.World);
        }
    }
}
```

Mapping with Keyword Manager, Keyword and actions

Our Keyword Manager, four defined keywords and the action for each keyword is ready.
Let's put them together.

1. Select the **Keyword Manager** component in the **Inspector** window. By default,
   Keywords and Responses Size is 0.

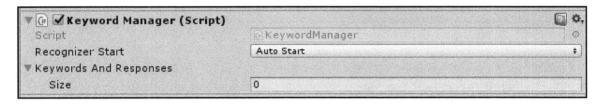

Keyword Manager script in inspector window

2. Update **Keywords and Responses Size** to 4 and press the *Tab* key. The property
   element for 4 different keywords will be created.

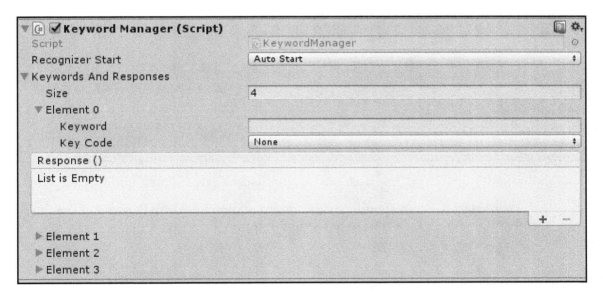

Response item for the action

3. Let's apply the first Keyword - **Show Speaker**. Then click on the (+) icon in the Response List.

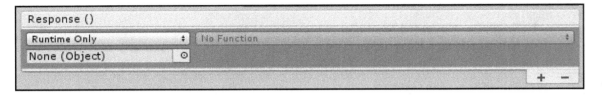

Adding Response List

4. In the response item, no object is specified. We want to apply the Voice Command to **Left_Speaker_geo** object. So, either drag and drop the object here from **Object Hierarchy** or click on the Object Icon and take the reference of the same speaker object.

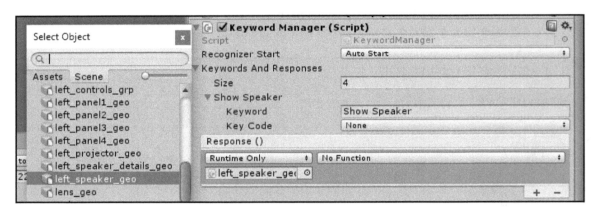

Linking the object with Keyword Manager

5. Then select the corresponding function from **Function Drop Down**. For the `ShowSpeaker()` we need to navigate through `VoiceCommandHandler | ShowSpaker()`.

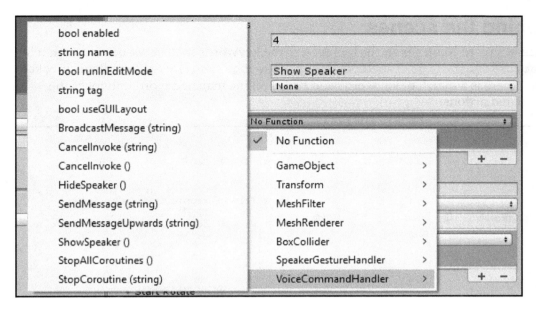

Add VoiceCommandHandler

You have successfully added your first Voice Command.

Apply the other three keywords and map the function respectively to the methods.

 Just keep in mind, for Rotate, you need to give reference of the entire HoloLens object as we want to rotate the entire object.

Voice Command – see it in action

Build, execute and run. When the application is running, you can say the following commands

- Show Speaker
- Hide Speaker
- Start Rotate
- Stop Rotate

You will see the command execution in action.

Behind the scenes

Once Keyword Manager starts, it processes the keywords that are associated with it in the `ProcessKeyBindings()` method. `OnPharseRecognized()` event handler is invoked when there is a match of keywords and based on the matched word it invokes the associated actions.

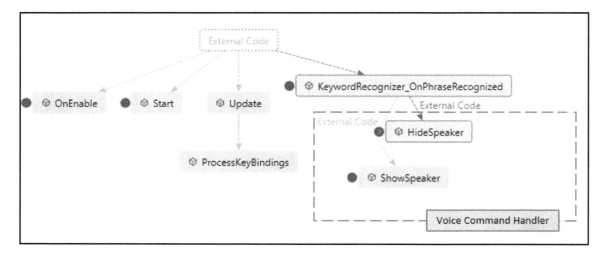

Keyword Manager and Command Handler in Code Map

Interaction using text to speech

Text to speech capabilities for your holographic applications bring another level of interactivity with virtual objects. You can place text to speech voices around your holograms to make your applications more collaborative. You can achieve it very easily with the help of the `TextToSpeech` manager available with HoloLens Toolkit.

Setting up text to speech Manager

Follow the steps for setting up text to speech manager:

1. Add an Empty Object in the **Object Hierarchy** option and name it **SoundManager.**
2. Select the **SoundManager** option from **Object Hierarchy**. Right-click on **SoundManager** and add new **Audio Source.**

3. Select the **Audio Source | Inspector** panel change the **Spatial Blend** option from **2D** to **3D**.

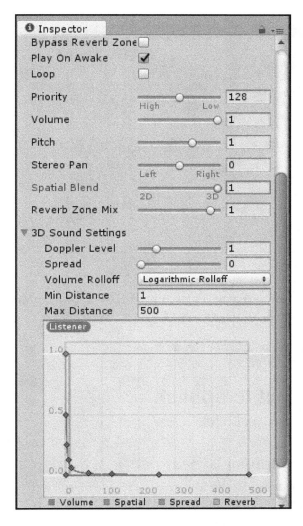

3D Sound Settings on Inspector windows.

4. Navigate to **HoloToolkit** |**Utilities** | **Scripts**, and drag and drop the **TextToSpeechManager** script over **SoundManager** as a component.

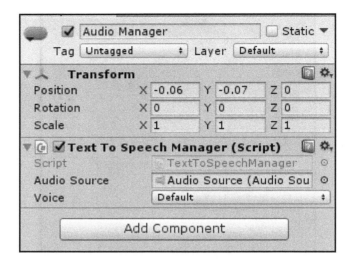

Adding Text to Speech Manager Component

5. Drag the **Audio Source** and attach it with the input parameter of **Audio Source** of **Text to Speech Manager Script**.

That's all for setting up the text to speech manager.

Integration of text to speech

Let's integrate this with our application now.

Text to speech in application launch

Add a new C# script file and add the following script for test TextToSpeechManager. We have named it SpeechText:

```
using HoloToolkit.Unity
public class SpeechText : MonoBehaviour {
    // Use this for initialization
    void Start () {
      TextToSpeechManager tt = this.GetComponent
            <TextToSpeechManager>();
      tt.Voice = TextToSpeechVoice.Mark;
      tt.SpeakText("Welcome To Explore HoloLens.
```

```
A Holographic View of a HoloLens device.
You can use Gaze, Gesture and Voice Command
to explore different components.
Walk around and start exploring !");


        }
    }
```

Finally, drag and drop this script to sound manager to execute first hand when the app is launched.

Text to speech in Air Tap

To add speech on Air Tap gesture, we can instantiate the TextToSpeechManager and call the SpeakText() method. To integrate speech while doing Air Tap on the Lenses, navigate to the LensGestureHandler class and add the following code snippet while tapped is true.

```
if (isTapped)
{
    ..........
    var soundManager= GameObject.FindWithTag("SoundManager");

    TextToSpeechManager tt =
        soundManager.GetComponent<TextToSpeechManager>();
    tt.SpeakText("The HoloLens display is basically a set of
        transparent lenses placed just in front of the eyes.");
}
```

TextToSpeechManager supports *four* different types of voices. You can choose a specific voice in your code from the *TextToSpeechVoice* enumeration.

Text to speech - see it in action

We integrate the speech to text in the app launch, so when your application launches, the Voice Command will be invoked. Along with that, when you perform an Air Tap on lenses, the speech command will also execute.

Behind the scenes

Consider the instance when you perform the Air Tap on Lenses, `OnInputClicked()` event invoke the `SpeakText()` method . Which calls `PlaySpeech()` from the `TextToSpeechManager` class to accept the text and execute it as voice. The `PlaySpeech()` method generates the speech stream and then plays it.

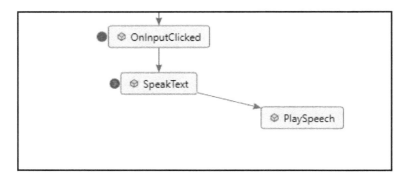

Text to Speech in Code Map

Deploying your app

We are now done with the development of our first holographic app. We have done it end-to-end, starting all the way from scenario identification to development of a 3D Model to scripting. It's time to see and experience a 3D application using device. There are several ways of deploying the application and they are listed as follows:

- Deploy using Visual Studio
- Deploy using HoloToolkit Build Window
- Deploy using Device Portal
- Deploy using Unity Remoting

Deploy using Visual Studio

Over the course of this chapter, we have already talked about the deployment of your holographic app using Visual Studio. Here, we generally build the app in Unity and take out the Visual Studio solution. Then, you open the solution from Visual Studio and run the app directly.

Deploy using Build Automation Window

The **Build Window** in HoloToolkit for Unity is extremely useful and speeds up the deployment. This window can be used for creating the Visual Studio solution, building the app, package creation, installing onto a device or an emulator by just one click. This utility makes the Unity3D and Visual Studio integration workflow faster.

Build windows of HoloToolkit

When we build the solution using HoloToolkit **Build Window**, it patches all the required assembly from Unity—these steps are done in several stages, then it proceeds for the creating Visual Studio Solution, and finally it creates the packages.

Once the package creation is done, you deploy it either to emulator or device for test it out.

This dialog window also provides several utilities such as Opening the Device Portal, Launching the App, Viewing the Log File and Uninstalling application.

Deploy using Device Portal

We can upload and install an `appx` package generated by Unity to device or emulator using the Device Portal. Once your Package for `ExploreHoloLens` app is created, you can upload the package in Device Portal.

Open the Device Portal from the left-hand side panel, navigate to **System | Apps**.

The **Installed Apps** section will have listed all the installed apps. We can select them to run it or remove them.

The **Running Apps** section shows the list of currently running app in emulator or in device. We can remove any **Running App** from this section by clicking on the cross icon associated with each **Running App**.

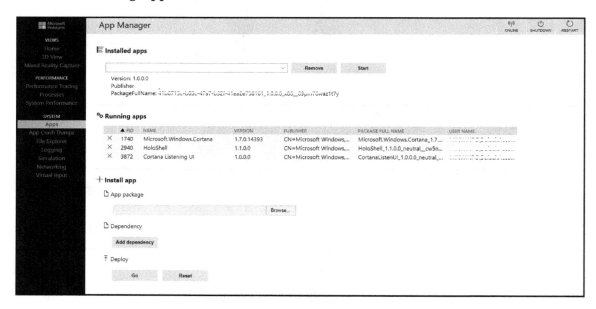

Apps screen of Device Portal

Load the package

To load and run an existing package, browse the app package for the `ExploreHoloLens` app. For an existing package, select the appx bundle or appx file from the browse dialog option and then click on Go. This will load the package and display a Done! message when it is successful. It will also display an error message if there are errors.

Once done, go back to the **Installed Apps** dropdown—You will have your newly loaded app listed.

Start the app

If you wish to start the app, select the app from the **Installed App** dropdown, click on start.

The emulator or device will load the newly selected app and run it, which eventually will also appear in the **Running Apps** list.

Deploy using Unity remoting

Holographic remoting reduces the development iteration round trip during deployment of holographic apps from Unity. When we are using holographic remoting to run an app, you don't need to build a project separately using universal windows apps or even need to run it from Visual Studio. In this case, you need to establish a connection between HoloLens and Unity. Once the connection is established, holographic remoting enables you to run your apps directly from the Unity editor.

Configure HoloLens device for remoting

To use this feature, you need to have a holographic remoting Player installed on a HoloLens Device. Follow the following steps to configure holographic remoting:

1. Install the app from Windows Store into your HoloLens device.

2. Launch the app, you will see the **Holographic Remoting Player** launched and display the **IP address** to connect with. This IP address is nothing but your HoloLens device IP Address.

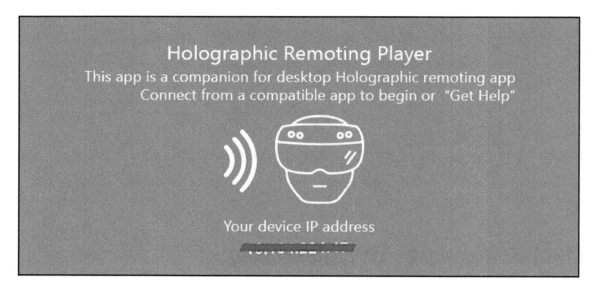

View Device IP Address on the Holographic Remoting player application

Holographic emulation – remoting from Unity

Once the installation is done and you see the IP Address in the holographic remoting Tool, it means your HoloLens device is ready to connect. Open the holographic app project in Unity and navigate to **Windows| Holographic emulation**.

This will launch the Holographic Emulation Window. In the holographic emulation window, Set the **Emulation Mode** to **Remote to Device** and provide the **Remote Machine IP address**.

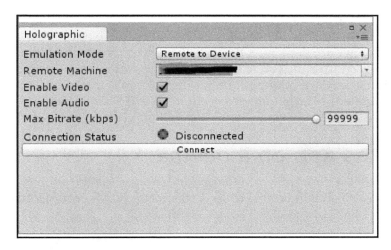

Set emulation mode-- Remote to Device and set device IP Address

 Make sure your development PC and HoloLens are in the same network or the IP address is accessible to each other.

Click on **Connect**. **Connection Status** will change from **Disconnected** I **Connecting**I**Connected.**

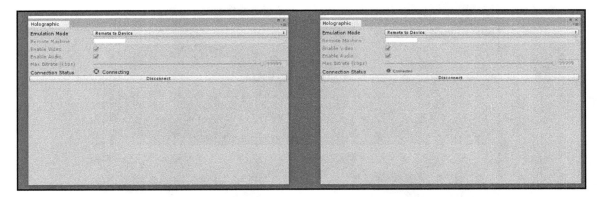

Connection state change from Connecting to Connected

That's it. You can run the app in Unity, and see it running on your device as well.

 When we are deploying through the holographic remoting process, the Unity editor runs you're the holographic content in the editor itself. Holographic remoting processes the holographic streams content from the editor to the development system to your HoloLens Device.

Running the app on a device

Once the deployment is done, now run the app in Unity or use the published app in your Hololens to run it. You should be able to see a room-size 3D Model of HoloLens. Go ahead and perform Gaze, Air Tap and other Voice Commands!

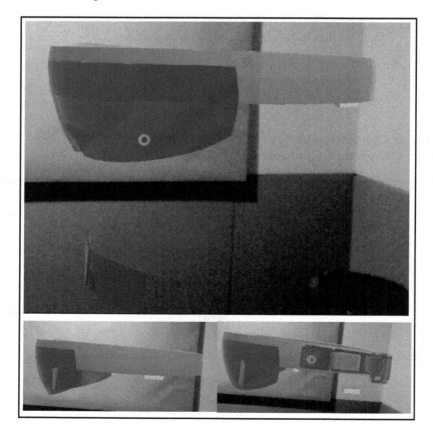

Running application in device

Testing your application

A holographic application is quite similar to testing a Windows application from some perspectives and is also very different from other perspectives. It's similar from the perspective of functional testing, performance testing, security testing and other similar factors, which we apply for any other Windows application testing. But testing of a holographic application is also very different from other Windows applications in factors such as lighting, sound, space, angles, and so on. These factors are very important, because holograms interact with real environments, and real space varies from one place to another. So, holograms should be designed and make it intractable to support variant real environments. Let us look at these factors in details.

Lighting conditions

Light can play a very important role while viewing the hologram. There could be a scenario which could make the whole experience bad, such as bright light coming from the back of the hologram, or from the side. Test your application in different lighting conditions. Also, test your application with other objects in space, like glasses or pitch black surfaces or transparent or reflective surfaces.

Space

Space is one factor which is always variable, from one room to another. One room could be very small with lots of furniture, and another room could be a very large one with lot of empty spaces. So, test your application in different spaces and your application should transition well between them.

Angle of views

Holograms are something which a user is viewing in a 3D Space, and they look real to the user. Users like to see real objects from different angles. So, test your application from different angles, like from sides, below, above, and even while walking inside your hologram.

View distance

Holograms are places in a 3D Space, and don't move along with the user. So, what happens if the user walks far away from the hologram or moves too close to the hologram, can the user still interact with the hologram? Test your application from these perspectives also.

Audio or visual hints

Again, always keep in mind that holograms are places in a 3D Space. What happens if the user places a hologram at a place and forgets about it, or if a scene wants the user to interact with the hologram, which is not in the current view area of user. For these scenarios, it's recommended to add either visual indicators or audio clues, so that the application can guide the user to the destined hologram.

Ambient sounds

Different levels of ambient sounds can interact with your Voice Commands. If the application you are designing has some Voice Commands, then remember to test them with different levels of ambient sounds.

Summary

This chapter explored the end-to-end development aspect of a holographic application. To build a holographic app, it is essential to understand the basics of Unity editor and different interaction models for a holographic app. In this chapter, we started the project by creating an empty Unity project and delved into different interactions by adding several scripts. We learned about HoloToolkit, and understand how easy it is to use for building a holographic application. This chapter also covered the various aspects of Gaze, Gesture, Voice Command, along with speech to text. The end of the chapter talked about different aspects of deployment of holographic apps in emulator, as well as in HoloLens Device. Then, we also discussed the different aspects of covering the testing scenarios of HoloLens. Overall, Chapter 3, *Explore HoloLens as Hologram - Scenario Identification and Sketching* and Chapter 4, *Explore HoloLens as Hologram - Developing Application and Deploying on Device* together, complete our trip of building our first project. Starting from envisioning, to deploying an app on to a device. This will give you a good start to building any standard holographic application. We will take this learning further and in the next chapter we will build a more complex series of applications using HoloLens.

5
Remote Monitoring of Smart Building(s) Using HoloLens - Scenario Identification and Sketching

In the previous chapter, we have successfully developed and deployed our first holographic application on HoloLens and tested it out. You have learned the complete cycle of holographic application development, starting from envisioning, scenario identification, sketching, 3D Modeling, setting up a development environment, gestures and interactions implementation using scripting, and, finally, deploying the application on HoloLens and testing it in the real world. The application we have developed and tested is an independent application, and doesn't interact with any other external systems. However, that will not be the case in real-world enterprise scenarios, where applications usually share data with other systems, either in real time or near real time, or in an offline mode.

In this chapter and the next chapter, you will learn to build the first connected holographic application that will pull data from an external system in real time, display it within your holographic application, and interact with it. For this application development, we will follow the familiar journey of holographic application development by doing the following things:

- Identifying a new scenario or application requirements
- Detailing that scenario through a sketching and planning phase?
- Creating 3D Assets

- Developing a scene in Unity3D and applying scripts
- Connecting this application with a backend service, pulling real-time data from a service, and updating holograms based on data received from external service
- Finally, deploying it on HoloLens device and testing it with near real time data

This journey of connected holographic application development is again divided into two chapters, this current one and the next one. In this chapter, you will learn about the following things:

- Envisioning
- Scenario prioritization
- Sketching
- Connected solution overview
- 3D Model overview

In the following chapter, you will continue this same scenario, and do the development and deployment on a HoloLens device.

Envisioning

As we considered a real-world example in the previous project, let's follow a similar process and make this application development close to realistic. Let's assume you are working for XYZ Inc., which is a leading building-management company responsible for managing the day-to-day maintenance and safety of various buildings spread out across different cities as well as countries. The maintenance and safety of these buildings is a very time consuming and tedious job. To simplify this job and reduce human errors, XYZ Inc. is planning to leverage automation by bringing new technologies, such as the **Internet of Things (IoT)**, connected solution automated alert systems, predictive maintenance, and very advanced real-time graphical views for administrators so that they can make decisions quickly and reduce the overall time for decision making.

To achieve this vision of XYZ Inc., the first thing you will do is to understand their vision better, and, to do that, schedule a meeting with the company leaders and stakeholders. In this meeting, you will get to know about their vision and requirements in detail. You can use familiar brainstorming ideas, such as a sticky notes session, to discussion requirements and do prioritization of those.

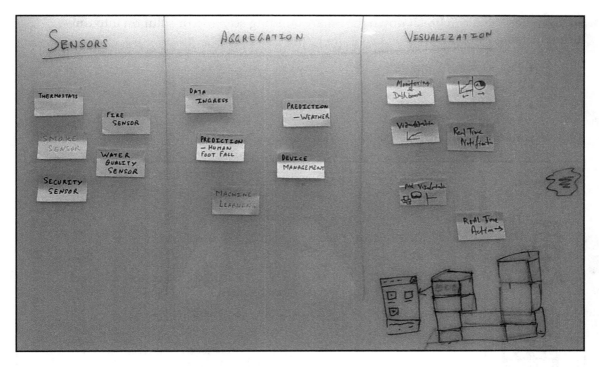

Brainstorming ideas for the Smart Building and its visualization

Solution scope overview

From the brainstorming session with customer stakeholders, you got to know about customer's vision for the Smart Building. Through the Smart Building vision, a customer wants to bring automation in to the day-to-day activities of buildings, such as **heating, ventilation**, and **air-conditioning (HVAC)**; lighting; security; water supply; and other activities.

Now, to achieve this vision, you need a set of connected devices; a centralized processing system, which can connect with these thousands of devices spread across geographies and manage these devices; and good analysis and visualization tools.

So, from here, you can break down the Smart Building scenario or solution into three different categories:

- Devices and Sensors
- Event Processing
- Visualization

Let's discuss these categories in detail:

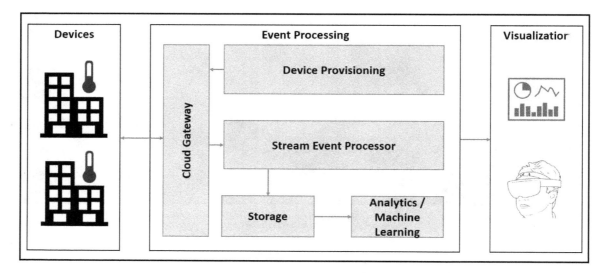

Solution overview

Devices and Sensors

Here, the customer has identified a list of sensors, such as thermostats, fire sensors, smoke sensors, security sensors, water-quality-check sensors, and others, which are part of almost every building that is managed by XYZ Inc. on day to day basis. These sensors need to be connected to a centralized Event Processing system to continuously process the data generated by these sensors.

Event Processing

Event Processing will be a centralized solution, to which all Smart Building sensors will connect and push data to at regular intervals. The Event Processing section can be further segregated into the following different sections, as per the different functionalities:

- **Cloud Gateway**: This is responsible for receiving a high data ingress from a large number of devices; it will also take care of sending commands to devices, such as software or firmware updates, and checking device states
- **Device Provisioning**: This is responsible for device on-boarding, managing, and de-provisioning devices on the platform
- **Stream Event Processor**: This is responsible for analyzing the stream of data received by the ingress layer and process stream, and then pushes it to storage
- **Storage**: This takes care of the large data storage and continuously stores the telemetry data received by a large number of devices
- **Analytics and Machine Learning**: This continuously analyzes telemetry data received from devices, applies Machine Learning algorithms to the data, and helps analysts/administrators to predict malfunctions in advance and prevent incidents

Visualization

There could be different methods for visualization, such as through web or desktop dashboards, mobile applications, and others. XYZ Inc.'s leadership wants to get a step ahead of the competition and wants to include MR as one of the visualization options. On further discussion with the team, two possible scenarios are identified for MR visualization:

- **Real-time monitoring**: Monitoring of device health using HoloLens application
- **Analyse and Act**: Visualizing and analyzing data using a MR application, in case of an incident, and immediately acting from within the MR application itself

Real-time monitoring

In a real-time monitoring scenario, the user is expected to visualize each device state in real time in 3D Space, and take the appropriate action based on the device state. The application is expected to connect with the Event Processing layer, and pull latest device state and present it to user so that the user can take informed decision.

Analyse and Act

In an Analyse and Act scenario, the user is expected to visualize the real-time device health, along with predictions and recommendations based on Machine Learning algorithms. Based on those recommendations, the user should be able to make an informed decision, and send a command or response back to the device through the holographic application itself. So, this scenario completes the user experience from the perspective of the administration of Smart Buildings.

Scenario prioritization

After identifying the scenarios, the next exercise is to prioritize scenarios. The selected prioritized scenario will be taken on to the next phase of the development. The team can either do a prioritization exercise in the same meeting or in another follow-up meeting. So, let's go through this prioritization exercise. During a prioritization exercise, the team discusses these scenarios from the perspective of immediate needs and dependencies. After reviewing both identified scenarios for *real-time monitoring* and *Analyse and Act*, they list the following two factors to be considered for the prioritization exercise:

- **Immediate need**: *Real-time monitoring* is required immediately, as that's the basic feature for an administrator to get familiar with device and application
- **Dependency** - *Analyse and Act* can start happening only after *real-time* monitoring has already been implemented

So, looking at the immediate needs and dependency, the team decided to prioritize *real-time monitoring* scenario over *Analyse and Act* scenario.

The next exercise is to elaborate on the prioritized scenario and start defining the interactions detail within this scenario.

Scenario elaboration

After the scenario prioritization exercise, the next activity is to elaborate on the selected scenario. In the elaboration, you will detail the features and user interactions within the scenario.

The objective of the selected scenario is to provide *real-time monitoring* of the devices across buildings. This means that the holographic application should be able to receive near--real-time data from devices. Let's first understand the IoT system architecture using Azure Services, and then we will elaborate on the selected scenario of *real-time monitoring* for visualization.

 The next section will provide a quick overview of IoT reference architecture for Remote Monitoring. If you are interested to learn it in detail, refer to Azure pre-configured solutions at `https://docs.microsoft.com/en-us/azure/iot-suite/iot-suite-remote-monitoring-sample-walkthrough`.

Internet of Things – Remote Monitoring

In this section, we will do a walk-through of the devices, event processing modules, and their integrations and data flow in detail. In the following section, you will learn about the integration of holographic application within event processing modules.

Within the event processing section/pillar, data flows through different sub-systems, such as the ingestion layer responsible for device connectivity and data ingestion, followed by the stream processing of data, and finally to the storage layer. Within the Azure IoT solution, a data ingestion layer is implemented using Azure IoT Hub and stream processing is implemented using Azure Stream Analytics.

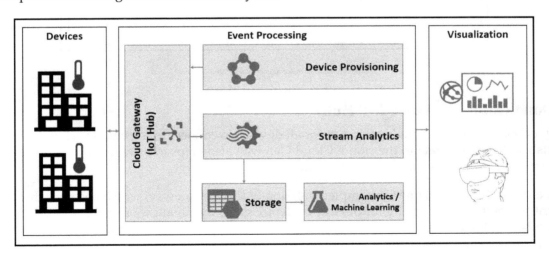

Solution architecture

Devices

Devices refers to actual physical devices, such as thermostats, fire sensors, smoke sensors, security sensors, water-quality-check sensors, and others. Depending on the type of device, they are split into two different categories:

- **IP enabled devices**: Devices that can connect over the Internet and can connect directly with a Cloud Gateway, such as Azure IoT Hub.
- **Low-powered devices**: Small low-powered devices are devices that connect on **Bluetooth low energy** (**Bluetooth LE** or **BLE**), which can't connect directly with a Cloud Gateway. So, for these devices, we will require a component such as device gateway, which connects with these low-powered devices, aggregates and consolidates data from these devices, and then send it across to the Cloud Gateway.

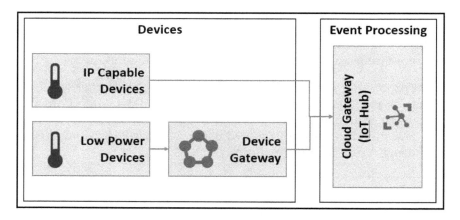

Type of devices and their connectivity with Cloud Gateway

Cloud Gateway (Azure IoT Hub)

Cloud Gateway is the single endpoint for all devices to connect to the cloud system and push all telemetry data to the cloud. Cloud Gateway also helps to provide bi-directional communication with backend systems and storage.

The Azure IoT Hub is one such implementation of Cloud Gateway, which becomes the central point of connectivity for devices or device gateways. It allows for a very large number of device connections and supports a huge volume of ingestion of telemetry to the cloud's backend. Also, it allows command flow to the devices and a mechanism to deliver commands to each individual device. It's also responsible for authentication and the authorization of each message coming in from devices.

Steam Processing (Azure Stream Analytics)

After the data has been ingested within the Cloud Gateway, the Steam Processing engine receives that telemetry and routes it to different down stream systems. This routing of telemetry messages can happen based on the type of message; for example, a particular type of message can be routed to a designated downstream system which processes only that particular type of message. This routing is very helpful in the case of segregating types of messages or taking priority action on a high priority message.

Azure Stream Analytics is one such Steam Processing engine available within Microsoft Azure, which allows the processing of data streams, such as aggregation of data, and then routing this stream of data to different permanent storage options such as Azure Blob, Azure Tables, Azure SQL Database, Azure Data Lake, and other data stores.

Data Store (Azure Storage)

After the data is stream processed, it is stored within a persistent data store, either as raw data received from devices, or in the form of processed data. Within Azure, there are multiple options available for storage, such as Azure Storage, Azure SQL Database, Azure Data Lake, and Azure CosmosDB. Each type of storage has a different purpose, such as Azure SQL Database is for storing and querying relational data, Azure Data Lake is a big data storage solution that provides a data analytics capability, and Azure CosmosDB is a schema--agnostic database engine. Out of this choice of storage services, we will use Azure CosmosDB DocumentDB API to store data in the form of schemaless JSON documents. Also, in the context of the current solution, we will store one document for each building, which will have details for all sensors within the building in one place.

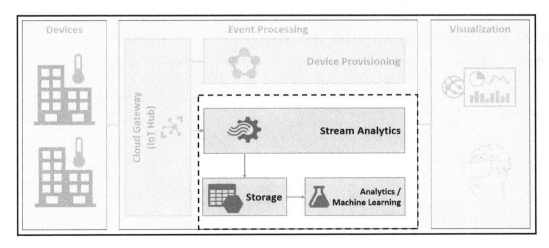

Highlighted Stream Analytics, Storage, and Machine Learning sections

Machine Learning

Machine Learning is a process for learning from existing data, and applying that learning to the current stream of telemetry data and predicting future outcomes. Azure Machine Learning is a cloud-based predictive analysis service, which can be plugged into the current solution, by creating a Machine Learning algorithm, training the model, and then applying it to the new upcoming data stream to have predictability in place. For the current scope of the project, we will skip the integration with Machine Learning, and will focus on integration with the holographic application.

To learn more about these technologies, refer to their documentation:
Azure IoT
Hub: https://azure.microsoft.com/en-us/services/iot-hub/
Azure Stream
Analytics: https://azure.microsoft.com/en-us/services/stream-analytics/
Azure Storage: https://docs.microsoft.com/en-us/azure/storage/
Azure SQL
Database: https://azure.microsoft.com/en-us/services/sql-database
Azure Data
Lake: https://azure.microsoft.com/en-us/solutions/data-lake/
Azure Machine
Learning: https://azure.microsoft.com/en-us/services/machine-learning/
Azure CosmosDB: https://docs.microsoft.com/en-us/azure/cosmos-db/

Real-time visualization through HoloLens

In the preceding section, we have learned about the data ingress flow, where devices connect with the IoT Hub, and stream analytics processes the stream of data and pushes it to storage. Now, in this section, let's discuss how this stored data will be consumed for data visualization within holographic application.

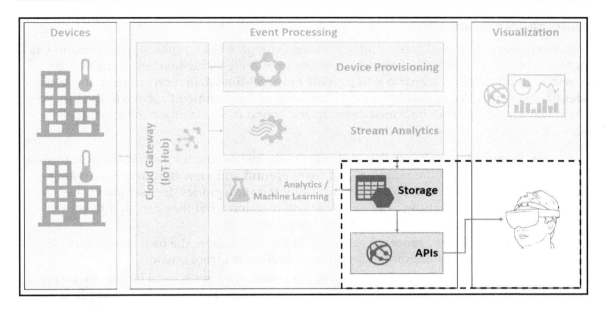

Solution to consume data through services

APIs

The holographic application is a specialized Universal Windows Application, so it can connect with any of the cloud services; all it needs is an Internet connection. So, to achieve this, we plan to create custom Web APIs, which will be consumed by the holographic application. These Web APIs will be responsible for exposing storage data--received from devices--and transforming and structuring the data in the required schema format as expected by the holographic application.

Holographic application

Visualization for Remote Monitoring is planned through a holographic application, and as per the scenario prioritization exercise, *real-time monitoring* is the finalized scenario. The objective of the selected scenario is to provide new real-time data received from the sensors, identify abnormalities or predict critical situations, and recommend actions to the user so that the user can make an informed decision and action it. To elaborate the scenario, let's break it down into following sections:

- **Primary View**: When the application is launched, the user should be able to view a hologram of the building, and roam around and view it from different angles. At the launch of the application, audio narration guides the user about the application features. The narrative will mention what they can all interact with within this application and how they interact.
- **Wings / floors / rooms**: At the start of the application, the user should be able to view different building wings, floors, and rooms from outside.
- **Sensors data**: On gazing at any of the rooms, the user should be able to see the last data received from the sensors for that room. For example, this application should show the last known temperature, fire warning state, and smoke warning state for any of the selected rooms within the building. On Air-Tapping on the room, the application will open a flyout with the room sensor details. The room details should contain the room location, such as wing/floor/room number, and so on, along with the sensor details, such as room temperature, smoke, and fire sensor details.
- **Fault details and recommendations**: In case there is a sensor fault or warning about high temperature, smoke, or fire, the holographic application will automatically get highlighted at the start of the application itself.

Sketching the scenarios

The next step after elaborating on the scenario details is to come up with sketches for the Remote Monitoring Smart Building scenario. As we discussed in the previous project's sketching phase, there is a two-fold purpose for sketching; first, it will be input to the next phase of asset development for the 3D Artist, and it will help to validate the requirements from the customer, so there are no surprises at the time of delivery. For sketching, either the designer can make it up on its own and build sketches, or they can get help from the 3D Artist.

Let's start with the sketch for the primary view of the scenario, where the user is viewing the building hologram to do the following:

- Roam around the hologram to view it from different angles
- Look at different interactive components

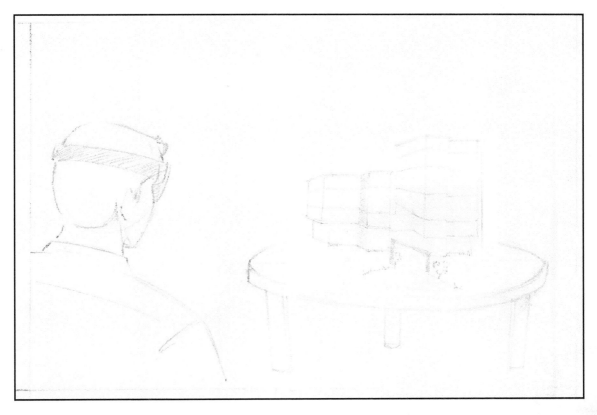

Sketch for user viewing the hologram for the building

Sketching - Interaction with Rooms and Sensors

While viewing the building hologram, the user can gaze at different interactive components; one such component is the room. When looking at any of the rooms, it should get highlighted and show the sensors' states associated with that room.

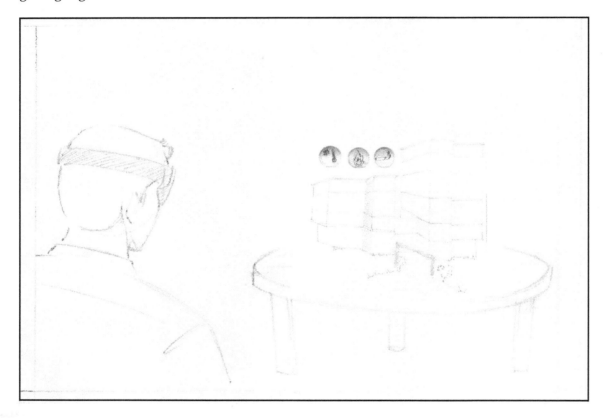

View sensor state on the gaze on the room

The user can then Air-Tap the room; this action should open up a flyout and the user should be able to view room details in detail.

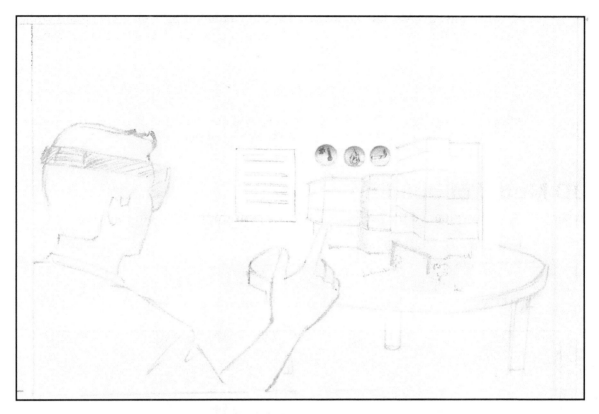

Air Tap on the sensor and get to view textual information and recommendation

3D Model – structure and components

In the preceding section, you have identified scenarios and detailed them through the sketching process. The next step is to develop the 3D Asset for the building and sensors that you are going to use in the next chapter to develop Remote Monitoring holographic application.

While developing our previous project, we have already gone through the detailed process of developing 3D Models. So, you have two options:

- **Create a new 3D Model**: Apply the 3D Model creation process you learned while developing the previous project, and develop a new 3D Model for the building and sensors

Or

- Download a pre-built 3D Asset using `https://github.com/hololensblueprint /HoloLensBlueprint/tree/master/Project2/Assets`

3D Model of building

After you have downloaded the pre-built 3D Asset, perform the following steps:

1. Create a new Unity3D project.
2. Import the asset to the Unity3D project.
3. Drag and drop the `office_building_fbx.fbm` asset within the scene.
4. You will see the following 3D Model of the building the scene area:

3D Model of the building

Now, open up the object of the building and visualize its hierarchy:

```
▼ office_building_fbx
   ▶ assembly_point_grp
   ▶ entrance_grp
   ▶ ground_grp
   ▼ icons_grp
        fire_off
        fire_on
        smoke_off
        smoke_on
        temperature_off
        temperature_on
   ▼ l_wing_grp
        l_wing_fifth_floor_glass
        l_wing_fifth_floor_wall
        l_wing_fourth_floor_glass
        l_wing_fourth_floor_wall
        l_wing_frist_floor_glass
        l_wing_frist_floor_wall
        l_wing_ground_floor_glass
        l_wing_ground_floor_wall
        l_wing_roof
        l_wing_second_floor_glass
        l_wing_second_floor_wall
        l_wing_third_floor_glass
        l_wing_third_floor_wall
        l_wing_top_wall
   ▶ lobby_grp
   ▶ r_wing_grp
```

Object Hierarchy for the 3D Model of the building

Within this hierarchy, you will observe that the Object Hierarchy is segregated into the two wings--the left and right wing groups--lobby, ground, entrance, and sensor icon groups. Also, expand the wing object, and observe that it consists of child objects for each floor and roof.

3D Model of sensors

After you have reviewed the building 3D Model, let's also review the sensors 3D Model. Within the hierarchy, you will find an object called **icons_grp**, which represents the icons for the sensors. We are going to use these sensors for scripting and gesture implementations.

3D Model of the sensor

Summary

In this chapter, you have started the development process for creating the first connected holographic application, and have learned about development aspects such as envisioning, scenario identification, scenario prioritization, sketching, architecture, and solutioning for the Azure IoT Remote Monitoring solution, and have received an overview of the 3D Model for the Smart Building.

In the next chapter, you will learn how to create a Unity3D project, import assets into Unity3D, script to implement gestures, build a complete solution, connect with Azure services, pull near-real-time sensor data, and deploy and test it on a HoloLens device.

6
Remote Monitoring of Smart Building(s) Using HoloLens - Developing Application and Deploying on Device

In the previous chapter, we began our journey by building a holographic application that leverages the **Internet of Things** (**IoT**) scenario. In this chapter, we will continue with the application development of the *Remote Monitoring of Smart Building(s)* that we identified throughout several exercises in the previous chapter. In the envisioning phase, we discussed the requirements and scenarios surrounding the identification for remotely monitoring a building, and talked about several scenarios such as temperature, fire, smoke detector, and monitoring the safety status of buildings or floors. As a part of the scenario's prioritization phase, we listed the scenarios that are to be taken forward for the development phase. We took a step further toward sketching to visualize and revalidate the identified scenarios, which illustrate a kind of vision demonstration of the app. We have also discussed the 3D holographic model of the Smart Building and its structure. Now that we have our clear development requirements, finalized scenarios, and 3D Assets ready to be used, we are all set to take the next step forward.

This chapter will cover all the steps from the setting up of the backend infrastructure, setting up your IoT environment, developing a holographic project, and connecting it with *Azure Services* till deploying it onto a device. A brief overview of the different aspects of this chapter are as follows:

- Setting up the backend infrastructure
- Building up the Azure IoT solutions
- Building a holographic application
- Connecting a holographic app with Azure
- Configuring, building, and testing a connected holographic app
- Extending the application

By the end of this chapter, you will have covered the complete end-to-end development process that requires building connected scenarios for a holographic application using several Azure IoT Services.

Solution development

This is going to be an IoT solution, which receives data from connected devices, stores the data and makes it available for consumption by the holographic application. So, in this section, you will first learn to set up the backend infrastructure--where you will have a simulated device--which connects with a Cloud Gateway. Data received by the Cloud Gateway is stored within persistent storage, and finally, this data will be made available to the holographic application through Web APIs.

Logical design diagram

The following is the logical design diagram for the overall solution, which is built on the same architecture design explained in the preceding chapter:

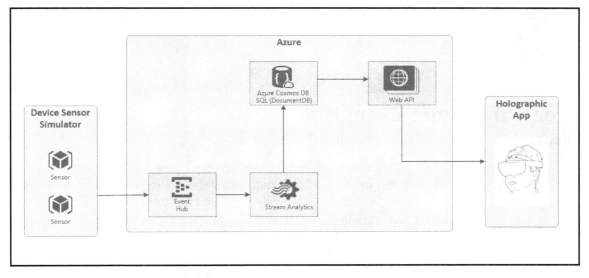

Logical architecture for remote monitoring of devices

 In this solution architecture, we use Event Hubs as a Cloud Gateway for event data ingress coming from the sensor. This is useful when there is only device-to-cloud data ingression flow. You can also use Azure IoT Hub; however, it is more appropriate when you have both device-to-cloud and cloud-to-device communication (bi-directional communication) flow, including device managements. You may refer to `https://docs.microso ft.com/en-us/azure/iot-hub/iot-hub-compare-event-hubs`, talks about the comparison of Azure IoT Hub and Azure Event Hubs.

Setting up the backend Infrastructure

To have infrastructure services, you will require the following Microsoft Azure Services:

1. Azure Event Hub
2. Azure Cosmos DB
3. Azure Stream Analytics

To start with, navigate to `https://portal.azure.com` and log in with your credentials. After that, perform the steps mentioned in the following section to create new services.

You have to make sure that you have a valid Azure subscription.

Create the Azure Event Hub

To create an Event Hub in Azure Portal, follow these steps:

1. From **New**, select the *Event Hubs* services and create a new Event Hub by providing a name, resource, and other required details, such as pricing tier or location.
2. Once the Event Hub is created and deployed, open the Event Hub and navigate to **Settings | Shared Access Polices**.
3. Create a **Policy** by selecting all claims, that is *Manage, Send, and Listen*.
4. Finally, note down its *connection string* and *event hub* name for further reference.

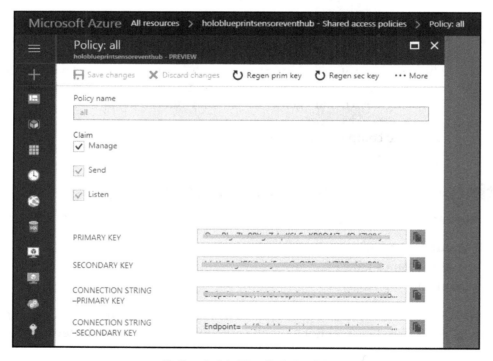

Fig: Connection String-Primary Key for Event Hub

Create the Azure Cosmos DB

The next step to set up our backend infrastructure is to create the Azure Cosmos DB; to do that, follow these steps:

1. From **New**, select **Azure Cosmos DB** service and create a new Cosmos DB by providing ID, resource group, location, and your subscription. You have to make sure that you select **SQL (DocumentDB)** from the *API Dropdown*:

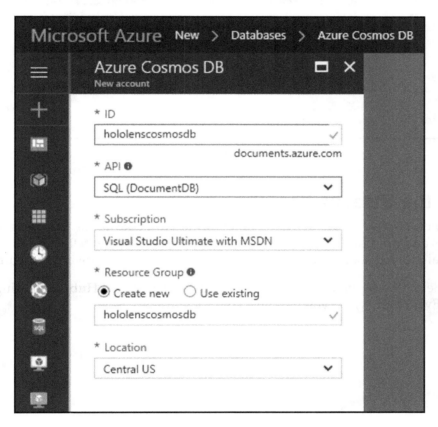

Create new Cosmos DB

2. Once "Azure Cosmos DB account" is created, add a collection by clicking on the "Add Collection" button and add a database with it.

3. Note down the **URI, PRIMARY KEY, PRIMARY COLLECTION STRING**, and **DATABASE CONNECTION STRING** names for further reference, which you can get from **Settings | Keys**.

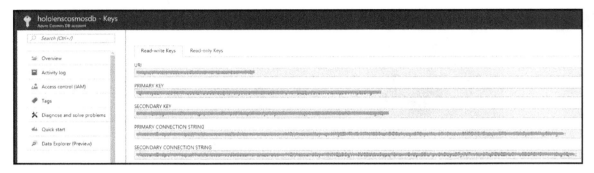

Taking reference of URI and Primary Key for Azure Cosmos DB

Create the Stream Analytics Job

Now, we should create a Stream Analytics Job, which will take up the data input from Event Hubs and process it in to Azure Cosmos DB--SQL (DocumentDB). Create a new Stream Analytics Job and, within the newly created Job, do the following configurations:

1. Select the **Input** tab and add a new input of type **Event Hub**, connecting to the Event Hub which you have just created above:

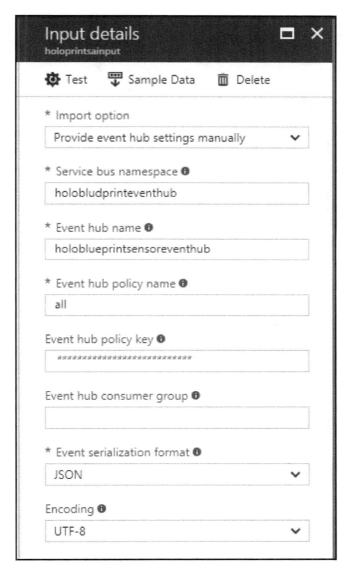

Input settings for Stream Analytics

2. Select the **Output** tab and add a new output of type *SQL-DocumentDB*, connecting with SQL-DocumentDB which you have just created above:

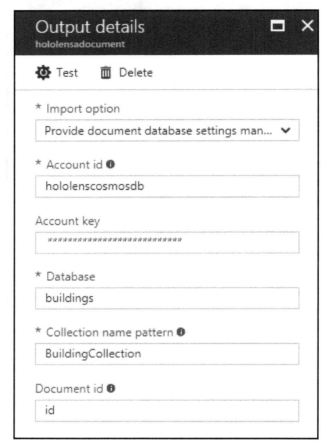

Output settings for Stream Analytics

3. Select a **query** tab, and add a simple query to connect the input and output stream:

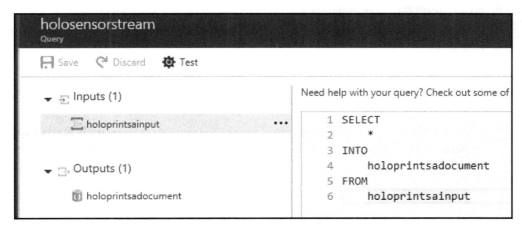

Query to connect input and output stream

Building up the Azure IoT solution

The next step is to create a simple simulator application to mimic a device to push data to the Cloud Gateway, that is, *Event Hub*. From there, data is pushed to *Azure Cosmos DB - SQL (DocumentDB)* using *Stream Analytics*, using the configuration we made in the preceding section. The second thing that we will create is the *Web API* application that will read data from Azure Cosmos DB - SQL (DocumentDB), and we will make it available to the holographic application for consumption.

Connecting Device with Azure Event Hub – simulator

1. Open Visual Studio and create a new console application using the context menu **File | New Project | Console Application**.
2. The next step is to add a NuGet Package `Microsoft.Azure.EventHubs` to this project.
3. To add this, right-click on the **Project** option with Solution Explorer, and select the context menu option, Manage NuGet Packages.

4. Now, search for the `Microsoft.Azure.EventHubs` NuGet package and add it to the solution.

5. Now, open the `Program.cs` file.

6. Add two constant strings to define the Event Hub connection string and even the hub name:

```
private const string EhConnectionString = "[Add connection
string]";
private const string EhEntityPath = "[Add event hub name]";
```

5. Add the following method `SendMessagesToEventHub()`, which reads random generated building sensor data, connects with Event Hub, and pushes sensor data to the Event Hub every second:

```
private static async Task SendMessagesToEventHub(int
numMessagesToSend)
{
    // Creates an EventHubsConnectionStringBuilder object from a
       the connection string, and sets the EntityPath.
    var connectionStringBuilder = new
        EventHubsConnectionStringBuilder(EhConnectionString)
    {
        EntityPath = EhEntityPath
    };
    var eventHubClient =
            EventHubClient.CreateFromConnectionString
                (connectionStringBuilder.ToString());
    // Load randomized building sensor data.
    Building building = BuildingCreator.LoadBuildingsData(0);
    for (var i = 0; i < numMessagesToSend; i++)
    {
        try
        {
            var message = $"Message {i}";
            Console.WriteLine($"Sending message: {message}");
            await eventHubClient.SendAsync(new
            EventData(Encoding.UTF8.GetBytes
            (Newtonsoft.Json.JsonConvert.SerializeObject(building)
        )));
        }
        catch (Exception exception)
        {
            Console.WriteLine($"{DateTime.Now} > Exception:
                    {exception.Message}");
        }
        await Task.Delay(1000);
```

```
    }
        Console.WriteLine($"{numMessagesToSend} messages sent.");
}
```

With this, our IoT solution is almost ready. The next thing we will focus on is creating the Web API for the consumer channel, in this case, our holographic app.

Developing the Web API

The Web API consumes the Azure Cosmos DB - SQL (DocumentDB) data received from sensors and exposes the REST API to be consumed by other applications. Here are the steps to follow to build the service:

1. Open *Visual Studio* and create a web application using the context menu **File | New Project | Web | ASP.NET Web Application (.NET Framework)**.

2. From the **New ASP.NET Web Application** dialog window, select the **Azure API App** option.

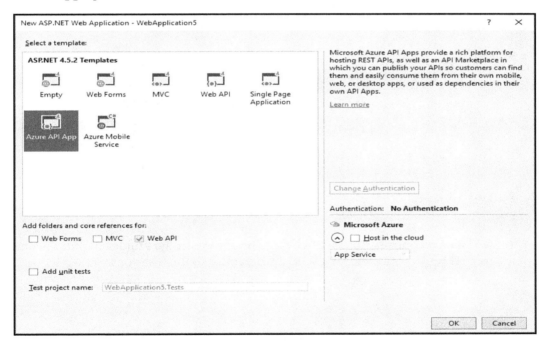

Azure API App selection

3. Within the project, add a new class, such as `BuildingCreator.cs`, which is responsible for connecting with SQL-DocumentDB and reading building sensor data.

4. Within the `BuildingCreator` class, add the following constant strings for connecting with SQL-DocumentDB:

```
private const string EndpointUri =
"https://[endpointname].documents.azure.com:443/";
private const string PrimaryKey = "[primary key]";
private const string DatabaseName = "[database name]";
private const string CollectionName = "[collection name]";
```

5. Now, within the BuildingCreator class, add a new method to fetch data from SQL-DocumentDB:

```
public Building LoadBuilding(string buildingNumber)
{
    this.client = new DocumentClient(new Uri(EndpointUri),
            PrimaryKey);
    FeedOptions queryOptions = new FeedOptions { MaxItemCount = -1
};

    // Here we find the Building via its Id/building number
    IQueryable<Building> buildingQuery =
            this.client.CreateDocumentQuery<Building>(
            UriFactory.CreateDocumentCollectionUri
            DatabaseName, CollectionName), queryOptions)
            .Where(f => f.Id == buildingNumber);
            for each (Building building in buildingQuery)
    {
        return building;
    }
    return null;
}
```

6. Add a new controller by right-clicking on the controllers folder within the newly created project and selecting **Add a Controller** within the context menu.

7. Within this controller class, add the following method:

```
public IHttpActionResult GetBuilding(int id)
{
    Building building = new
        BuildingCreator().LoadBuilding(id.ToString());
    if (building == null)
    {
        return NotFound();
    }
    return (Ok(building));
}
```

In the next step, deploy the services in Azure. There are several ways to do that. The easiest and fastest way that you can do it is from Visual Studio itself. Right-click on the **WebAPI** project and select **Publish**.

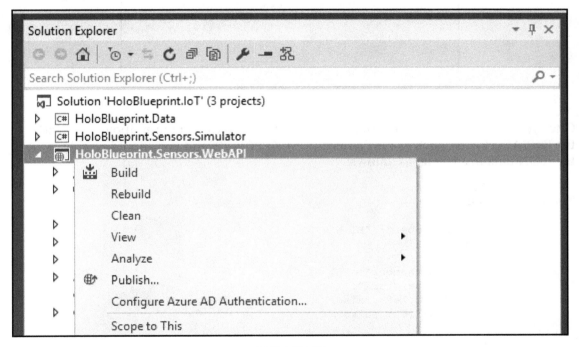

Publish the Web API

It will launch the Application Publish wizard, from where you can select the **Microsoft Azure App Service** option and follow the wizard steps to complete the deployment process.

Web API Publish Wizard

Once your service is hosted and deployed on Azure, you can hit the endpoints with any of the Rest API clients (such as the *Postman REST API Client*), and you should be able to view the JSON data in a format as follows:

```
"WingName": "B1-W1",
      "Floors": [
        {
        "FloorNumber": "B1-W1-F0",
        "Rooms": [
          {
            "RoomNumber": "B1-W1-001",
            "Temperature": 24,
            "IsSmoke": false,
            "IsFire": false
          }
        ]... ... ....
```

 Refer to `https://github.com/PacktPublishing/HoloLens-Blueprints` for a sample of this solution.

Setting up a project for a holographic application

The creation of a project, setting up the basic Unity Editor, and using the HoloToolkit are similar for all holographic applications. We have already discussed that in detail in Chapter 4, *Explore HoloLens as Hologram - Developing Application and Deploying on Device*, where we created our first holographic application—*Explore HoloLens*. In this project setup section, we will mostly cover the same things, but at a very high level. Anytime you need to check out the details, you can refer to Chapter 4, *Explore HoloLens as Hologram - Developing Application and Deploying on Device*.

Getting started – creating a new project

To start with, we will create a new project in Unity by following these steps:

1. Run a new instance of **Unity3D.**
2. Create a new **3D Project** by providing a project name and all other required details.
3. Click on **Create Project.**

Creating a Remote Monitoring project in Unity

This will launch the Unity Editor with a newly created project. By now, you must be familiar with this editor and several concepts of Unity that are required for application development. Here, we will jump directly into action.

Import the package

During the previous project, we saw the end-to-end process to create the package for HoloToolkit and use it in our Explore HoloLens application. We discussed that we will reuse the same code for all the projects, and here is the first reuse. As the next step of the development, we will first import the `HoloToolkit` package in to the Unity project.

> You can refer to the *Setting up the HoloToolkit-Unity* section in `Chapter 4`, *Explore HoloLens as Hologram - Developing Application and Deploying on Device*, for details of package creation.

In the Unity editor, right-click on the `Assets` folder, and from the context menu, navigate to **Import Package** | **Custom Package** and select the previously created `HoloToolkitUnityPackage`:

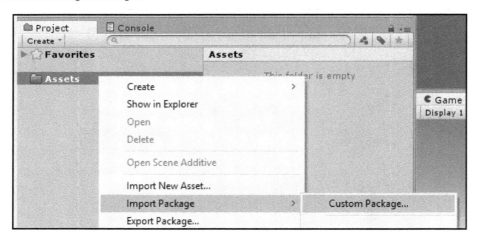

Importing HoloToolkit package

As a standard import process, the **Import Package** window will list all the scripts and components that are being imported. You do not need to change anything at this point in time; just click on **Import**.

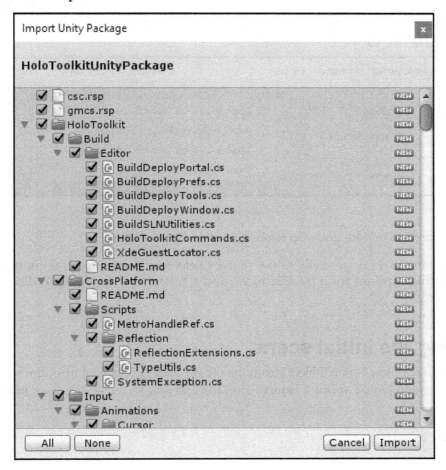

Importing Package in Unity

Within a couple of minutes, the import process will be completed, and you will have all your prerequisite assets and scripts imported in your Unity project.

Applying Project Settings

Once the HoloToolkit Package is installed, the HoloToolkit option will appear in the main menu item of the Unity Editor. Navigate to **HoloToolkit| Configure** and select **Apply HoloLens Scene Settings** and **Apply HoloLens Project Settings** to apply standard settings for the holographic application within editor.

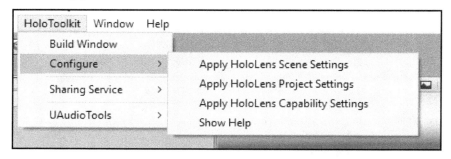

HoloLens menu options

Unity will ask for a reload, and you should go ahead and do that.

In case you have to do it manually rather than by selecting **Configure** options, navigate to **Build & Settings** options from the **File** menu and set the properties in the **Build Settings** window.

Updating the initial scene

Remove Main Camera Game Object (from the Object Hierarchy), and from the `Asset` folder, navigate to **HoloToolkit | Input | Prefabs** and add `HoloLensCamera` into the scene.

Importing 3D Assets – building model

Once we have the required objects present in the game scene, it is time to import the 3D objects of the building.

 You can download the FBX file of the 3D building for this project from the shared on `https://github.com/PacktPublishing/HoloLens-Blueprints`.

To import it into our project, right-click on **Assets** | **Import Assets** and select the FBX file from the download location:

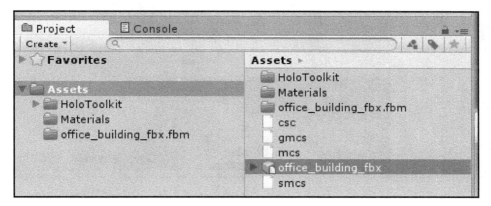

Import building FBX file

Once the import is done, drag and drop the `Office_bulding_fbx` object to **Object Hierarchy**:

- Scale up the building object to 2 for all the scale positions to get a bigger view, and that will give an excellent experience when you view through the device
- Select the `HoloLensCamera` object in **Object Hierarchy** and change the Y-Rotation point to -180 for the front view of the building

Then, you can adjust the distance between the camera and building model as per your want. With this, our initial setup of the Remote Monitoring Solution is ready, and this is how it should look in the Unity Scene View and Game View:

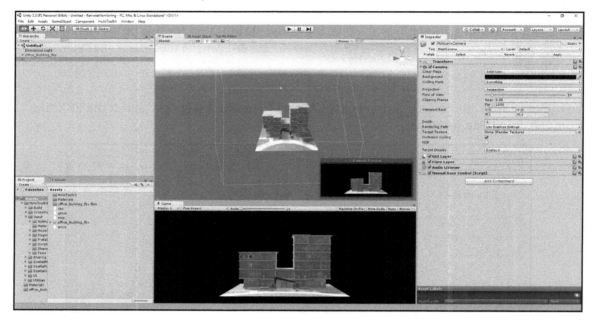

Initial View of the Scene

Exploring the building's 3D Objects

Once you import the 3D Model of the building, you can explore the hierarchy inside the **Object Hierarchy** browser. It consists of several smaller Game Objects, as you have seen in the previous chapter.

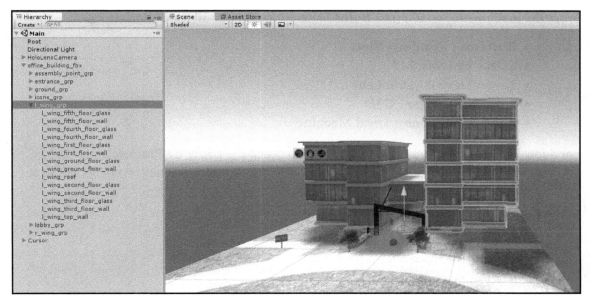

HoloLens 3D object in Object Hierarchy window

The building is divided into two wings with a set of floors. Each floor will have a sensor Game Object; at present, we have them for a single floor. However, using the script, we will make them work for other floors as well. For each floor, we can see there are two Game Objects, one for wall and one for glass. For the scripting purpose, we will use the Glass Game Object, which covers the entire floor from all sides.

Save the scene

After the basic setup and object insertion is done, let's go to **File | Save Scene** by providing a scene name, let's say, *Main*.

With that, we are all set with our project setup, and our holographic application is ready to build and run.

Build, run and test the initial setup

When our basic setup for the project is done, we can just build and run the application to test whether everything is working fine. In the earlier project development exercise chapter, we already talked about setting up the development environment and running a holographic app. Let's follow the same steps here.

From the main menu, go to **File** | **Build Settings** and verify all the build settings:

- **Platform** | **Windows Store**
- **SDK** | **Universal 10**
- **Target Device** | **HoloLens**
- **UWP Build Type** | **D3D**
- **Build and Run on** | **Local Machine**

For a local development reference, you can also check the **Unity C# Projects** and **Development Build** checkbox. Also, make sure that you add your main scene by clicking on the **Add Open Scenes** button.

In the **Build Settings** window, click on **Build**.

1. Create a **New Folder** named RemoteMonitoringApp.
2. Single **Click** on the RemoteMonitoringApp folder.
3. Press **Select Folder.**

From here, Unity will take it forward and perform the project builds and will create the solution for Visual Studio. When Unity is done, a **File Explorer** window will appear:

1. Navigate to the RemoteMonitoringApp folder and open the Visual Studio Solution created by Unity.
2. In Visual Studio, change the build Platform target from **ARM** to **X86**.
3. Select **HoloLens Emulator** from the list of run options.
4. From the main menu, click on **Debug** | **Start Without debugging** or just press *Ctrl + F5*.

Wait till the HoloLens Emulator starts and your application launches. Once the App is launched, you should be able to see the 3D Model of your hologram, as shown in the following image. If you are running it using the HoloLens device, you can experience a 3D building model and walk around it. Even if you see some of the sensor indicators in the building, they are not connected to the backend as of now, and at this point in time they are just part of the 3D building base model.

3D building

 You are also now familiar with different deployment mechanisms, and you have seen how to use them during the previous project; you are free to choose any of the processes for deployment and test it out.

All our initial setup is tested and ready. Let's start digging into a more scripting part of the holographic app.

Tracking building floors from 3D Model

In the next step of the development, we will enable the selection for each floor using Gaze; as per our design, we are receiving data based on each floor. So, let's enable the gaze for each floor.

Adding Gaze

When a user gazes at a floor, a cursor will indicate the gazed floor. To enable the gaze and have a cursor in your scene, we will use components from HoloToolkit as in the other exercise.

The following are the steps for that:

1. Add an empty Game Object in **Object Hierarchy** and rename it **Root.**
2. Go to the Assets folder, navigate to **HoloToolkit | Input| Scripts.**
3. Drag and drop the **InputManager** script to **Root** object.
4. Go to the Assets folder and navigate to **HoloToolkit | Input | Scripts | Gaze.**
5. Drag and drop the GazeManager and GazeStabilizer script to the **Root** object.

The GazeManager class manages everything related to gazing at an object based on the Raycast Hit. The GazeStabilizer class helps to stabilize the gaze. It samples the data form several *Raycast Hit* positions and helps in stabilizing your gaze for precision targeting. The InputManager class manages the sources of inputs.

Adding the cursor

A cursor is used to provide an indication to the object which is currently being gazed at. In this case, we will use the default cursor from the HoloToolkit. To add a cursor into the scene, perform the following steps:

1. Navigate to Assets | **HoloToolkit | Input | Prefabs | Cursor.**
2. Drag and drop the **Cursor prefabs** in the **Object Hierarchy.**

The script attached to the Cursor Game Object (*ObjectCursor*) takes care of everything along with interacting with GazeManager.

Adding the Box Collider

Once the GazeManager and cursor are in place, the next step is to identify the gazeable objects. As you know, the Box Collider is the base for collision primitives in Unity. Here, we will handle the gaze for each floor; hence, we would need to add a Box Collider component to each floor object.

For this 3D Model, we have two wings and a few floors in each wing (*right Wings – floor 0 to floor 3 and left wings – floor 0 to floor 5*), which you can clearly identify from the model. Select individual floor objects from the wings and add Box Collider components from the **Inspector** window. The following steps explain the process for one of the floors:

1. Expand the **r_wing_grp** Game Object from **Object Hierarchy**.
2. Select **r_wing_third_floor_glass**.
3. Navigate to the **Inspector** window | **Add Components** | **Add Box Collider**.

Box Collider added in right wing--3rd floor

Follow the same steps for all the floors and add the Box Collider for each of them so that we can start gazing at each floor.

Box Collider added in each floor

Add tag to each floor for mapping

The data we are receiving from the backend is mapped with each floor. So, we must identify for which floor we are receiving what data. To do that, we will first tag all floors as *<Building No>-<Wing No>-<Floor No>*. The following is a list of tags, which are mapped with each floor:

	B1-W2-F5 B1-W2-F4
B1-W1-F3	B1-W2-F3
B1-W1-F2	B1-W2-F2
B1-W1-F1	B1-W2-F1
B1-W1-F0	B1-W2-F0

Create the tag from the **Inspector** window and map them accordingly to each floor. The following screenshot shows one of the mappings of tags for *B1-W1-F3*:

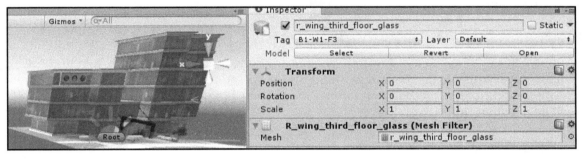

Adding tag for building floor

With that, our 3D Model has each floor uniquely defined, and we can start interacting with them.

Tracking Gaze for each floor

1. Create a folder called Scripts inside the Assets folder.
2. Add a new script by navigating from the context menu **Create | C# Scripts**, name it FloorInteractionManager.
3. Open the script file in **Visual Studio**.

By default, the `FloorInteractionManager` class is inherited from `MonoBehaviour`. `MonoBehaviour` is the base class from which every Unity script derives. In the next step, implement `IFocusable` interface in the `FloorInteractionManager` class. This interface implements the methods `OnFocusEnter()` and `OnFocusExit()`, which invoke when Gaze Enter and Gaze Exit, respectively. So, the default skeleton for the `FloorInteractionManager` class would be as follows:

```
public class FloorInteractionManager : MonoBehaviour, IFocusable {
    public void OnFocusEnter()
    {
    }

    public void OnFocusExit()
    {
    }

    // Use this for initialization
    void Start () {
}

    // Update is called once per frame
    void Update () {
    }
}
```

At this point in time, we don't want to change anything in the base skeleton. Just to check whether everything is tagged and working fine or not, add only a line inside the `OnFocusEnter()` method that returns the tag name of the Gazed object eventually that will be our floor number:

```
public void OnFocusEnter()
{
    var floorNo = this.gameObject.tag;
}
```

Attach the `FloorInteractionManager` script with all the floor glass objects where we have added the Box Collider. Once done, if you inspect any of the floors in the **Inspector** window, they will have the following components:

Floor Game Object attached with script

While *Transform, Mesh Renderer*, and *Mesh Filter* come with a base 3D Model, we have added the *Box Collider* and script as components as a part of the development.

That's it!

See Gaze in action

Save the scene and build and run the solution in the emulator once again. Now, you should be able to see the *cursor* for each *floor* as soon as it is gazed at.

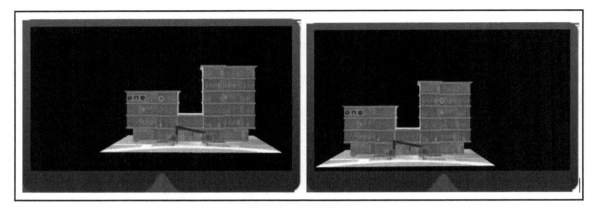

Gaze and Cursor on Floors

Verify the Gazed floor

If you look around the building--apart from where we intended to have the cursor by placing the Box Collider (for each floor)--you should see only a basic indication. However, the cursor will only be visible to those intended objects. That is how the different states of cursor work when it is on a gazeable object and when it is not.

To verify whether we can track the respective floor while it is being gazed at or not while in debug, follow these steps:

1. Open `FloorInteractionManager.cs` and place a breakpoint inside the `OnFocusEnter()` method.
2. Perform the Gaze in your holographic app.
3. As soon as you gaze at any of the floors, the breakpoint will be hit and the code execution will be paused.

At this time, if you explore the Game Object and its tag name, it will be the same associated tag, which was mapped with each floor.

```
 9   public void OnFocusEnter()        🔧 this.gameObject.tag  🔎 ▾ "B1-W1-F2"
10   {
11      var floorNo = this.gameObject.tag;
12   } ≤ 303ms elapsed       ◢ 🔧 this.gameObject {BaseObject: NotAvailableDuringDebugging} ⊕
13                               🔧 active                   true
                                 🔧 activeInHierarchy        true
     8 references               🔧 activeSelf               true
14   public void OnFocusExit()  ▷🔧 gameObject             {BaseObject: NotAvailableDuringDebugging}
15   {                           🔧 hideFlags               None
16                               🔧 isStatic                false
17   }                           🔧 layer                   0
18                              ▷⚙ m_CachedPtr             {318434752}
19   // Use this for initialization  🔧 name  🔍 ▾ "r_wing_second_floor_glass"
     0 references               ▷🔧 scene                  {UnityEngine.SceneManagement.Scene}
20   void Start () {             🔧 tag     🔍 ▾ "B1-W1-F2"                           📌
21                              ▷🔧 transform              {BaseObject: NotAvailableDuringDebugging}
22   }                          ▷⚙ Static members
                                ▷⚙ Non-Public members
```

Tracking the floor object

With this, we have our Gaze working. We can focus on any of the floors and, using the code, we get the reference of the floor name that we will use further to map the data coming from backend services.

Connecting a holographic app with Azure

In this section, we will connect our application with Azure, receive the data, and then bind the data Game Objects.

Defining data models

In the holographic app, the data model will hold the data returned by the *Building Rest Services*, hence it should be same structure. You can use any of the *Rest Client Applications* and hit the service end to get the JSON response from services, and based on that you can create the data model; for that, perform the following steps:

1. Add a new folder inside the `Script` folder called `Building Models`:
2. Add four different script files: `Building`, `Wing`, `Floor`, and `Room`

Define all the classes, as follows:

```
[Serializable]
public class Building
{
    public string BuildingName;
    public string Id;
    public string Address;
    public List<Wing> Wings;
}

[Serializable]
public class Wing
{
    public string WingName;
    public List<Floor> Floors;
}

[Serializable]
Public class Floor
{
    public string FloorNumber;
    public List<Room> Rooms;
}

[Serializable]
public class Room
{
    public string RoomNumber;
    public int Temperature;
    public bool IsSmoke;
    public bool IsFire;
}
```

Ensure that you add serializable attributes for each class so that it can be serialized.

This is how our Building data model looks with *Collection Association*:

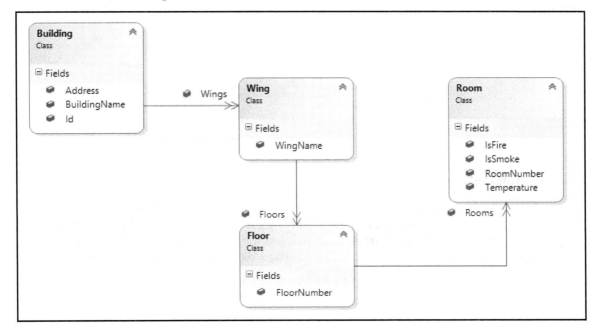

Building data model class diagram

 You may wonder why we have used data fields rather than using properties for the given data models. Unity cannot serialize properties; that's why all the members in the data model are defined as fields rather than properties. Refer to `http://docs.unity3d.com/ScriptReference/SerializeField.html` for more information related to Unity serialization.

Enabling the InternetClient capabilities

Providing access to the specific capabilities, we let our application allow specific hardware or external components. Enabling access to InternetClient, we allow our holographic application to interact with external web requests.

From the Unity Main Menu, go to **HoloToolkit** | **Configure** | **Apply HoloLens Capability Settings**:

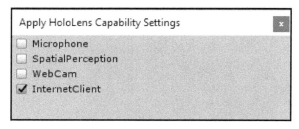

Apply HoloLens Capability Settings
☐ Microphone
☐ SpatialPerception
☐ WebCam
☑ InternetClient

Applying HoloLens capability settings

Check the **InternetClient** option to enable.

 You can also enable this setting by navigating to **Edit** | **Project Settings** | **Players and Under Settings for Windows Store** | **Publishing Settings** | **Capabilities**; you will find the **InternetClient** option. *Applying HoloLens Capabilities Settings* from the *HoloLens Toolkit* internally does the same settings only.

Building Azure Bridge

Now we will create an Azure Bridge, which fetches the data from the services and maps it with the *Building Data Model*, which we have just defined in the preceding section:

1. Navigate to the `Assets| Scripts` folder.
2. Add a new script by navigating to context menu | **Create** | **C# Scripts** and name it **AzureBridge**
3. Open the script file in **Visual Studio**

As in the first step, we will make this class a *Singleton* to use only one instance of it. You can achieve this by just inheriting the class from the *Singleton* class, by passing the same class type as parameter:

```
public class AzureBridge : Singleton<AzureBridge> {
    // Use this for initialization
    void Start () {
    }
    // Update is called once per frame
    void Update () {
    }
}
```

The background work for creating and managing instances is taken care of by the HoloToolkit utility. You can view the Singleton class definition and check what is happening inside, where it returns a new instance only if it is null, otherwise it returns the same object.

Getting the data

Connecting with the Web API and fetching the data to our application can be done in several ways. In this project, we will use UnityWebRequest to communicate with our backend services.

 UnityWebRequest allows Unity games to interact with backend services over HTTP Request and Response.

Once data gets downloaded by the DownloadHandler, we need to convert it from the JSON response to building model. The JsonUtility helps with that conversion.

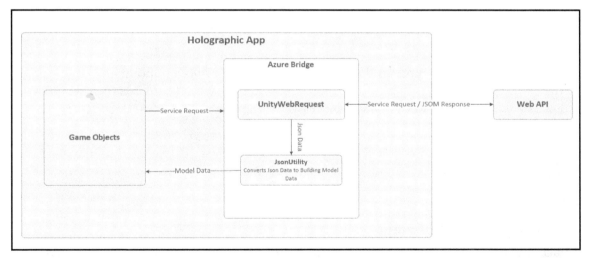

Azure Bridge communication

Here is the `AzureBridge` class implemented with UnityWebRequest:

```
public class AzureBridge : Singleton<AzureBridge>
{
    public Building buildingModel;
    void Start()
```

```
        {
            StartCoroutine(GetBuilding());
        }

    IEnumerator GetBuilding()
    {
        // Replace the URI with your hosted service URL
        UnityWebRequest webRequest =
            UnityWebRequest.Get("http://holoblueprintsensorswebapi.
            azurewebsites.net/api/building");
        yield return webRequest.Send();
        if (webRequest.isError)
        {
            Debug.Log(webRequest.error);
        }
    else
        {
            string sensorData =
                System.Text.Encoding.UTF8.GetString(webRequest.
                downloadHandler.data);
            buildingModel = JsonUtility.FromJson<Building>
                (sensorData);
        }
    }
}
```

The `GetBuilding()` method sends a `WebRequest` to the service endpoint and once data returns from the service, the `JsonUtility.FromJson()` method converts the raw *JSON* data into our desired building model objects.

 We call the `GetBuilding()` method within `StartCoroutine()`, which returns a coroutine. When you start a coroutine, it's executed until a yield instruction is found. Refer to `https://docs.unity3d.com/ScriptReferen ce/Coroutine.html` for more details.

Attaching the Azure Bridge

Attach this script with the Root object of our holographic app, by just dragging the script into the Root Object, either in **Object Hierarchy** or in the **Inspector** window.

With this, our *Azure Bridge* is ready to connect and fetch the live data. If you run the application by placing a breakpoint inside the `GetBuilding()` method, you should be able to explore the data retrieved from the services.

Binding data with sensor indicators

We have the data from backend services, and now onward our job is much simpler and straight forward. You need to find out the required object for the data we have received and set the values accordingly.

At this stage, we have data from temperature, fire, and smoke sensors. This data needs to be displayed against the Sensors Object that we have in the holographic app. If you refer to the *"Exploring the building's 3D Objects"* section, you will see the structure for *Icon_grp*, which consists of all the indicators for *temperature, smoke, and fire*.

Data binding with sensor indicators

Let's perform a few steps to make the **fire** indicator work with live data:

1. The fire indicator has two different icons: *fire_off* and *fire_on*.
2. Tag them as **FireOff** and **FireOn**, respectively.
3. From the Assets | Scripts folder, navigate to the FloorInteractionManager.cs file and open it inside *Visual Studio*.
4. Go back to the OnFocusEnter() method. So far, you have only one line there, which is referring to the *floorNo*.
5. Update the OnFocusEnter() method with the following line of code:

```
public void OnFocusEnter()
    {
        string floorNo = this.gameObject.tag;
        Floor floor = AzureBridge.Instance.bm.Wings.Select
          (item => item.Floors.Where(y => y.FloorNumber ==
          floorNo).FirstOrDefault()).FirstOrDefault();

        if (floor == null)
{
            return;
}
```

```
                     bool IsFire = floor.Rooms[0].IsFire;
                         this.SetFireIndicator(IsFire);
    }
```

The code is self-explanatory enough. `this.gameObject.tag` returns the *floorNo*, and we used that with a LINQ statement to get that particular floor's information. *AzureBridge.Instance* returns an instance of the *AzureBridge*, and `AzureBridge.Instance.bm` holds the values of the *Building Model* returned by the service. Considering we have a single room on each floor, `floor.Rooms[0].IsFire` set up the values of the fire sensor in the `isFire` local variable. Then, create a new method called `SetFireIndicator()` and pass the `IsFire` value as a parameter. Add the following code block inside the `SetFireIndicator()` method:

```
private void SetFireIndicator(bool isFireIndicator)
{
    GameObject fireonObject = GameObject.FindWithTag("FireOn");
    GameObject fireoffObject = GameObject.FindWithTag("FireOff");

    MeshRenderer fireonrenderer =
        fireonObject.GetComponentInChildren<MeshRenderer>();
    MeshRenderer fireoffRenderer =
        fireoffObject.GetComponentInChildren<MeshRenderer>();

    if (isFireIndicator)
    {
        fireonrenderer.enabled = true;
        fireoffRenderer.enabled = false;
    }
    else
    {
        fireonrenderer.enabled = false;
        fireoffRenderer.enabled = true;
    }
}
```

In this method, we took the references of fire indicator objects and enabled and disabled the `MeshRenderer` based on the value of `isFireIndicator`.

That's it. Now, if you run the application, you should be able to see the live data from services to fire.

Similar to what we have done for fire, we can do for smoke and temperature indicator objects. For fire and smoke indicators, we set the visibility based on either true or false; in the case of temperature, we will display the red indicator if it's more than 24, otherwise it will display blue for normal temperature. That concludes the end-to-end scenarios for the applications.

Instead of repeating the code for getting objects from tag name and finding `MeshRenderer` from the object for every single Game Object, we can create a generic method, passing the object that returns what we need.

See the live data in action

With all that binding complete for sensors indicators, build and run the application either in Emulator or in your device. You will find that the indicator changes based on the data received from Services.

Fig : Sensor Indicator Data Binding

The indicator on the left-hand side of the preceding two scenarios shows no fire, no smoke on the floor, and temperature as normal, whereas the right-hand side indicates a high temperature.

Enabling sensor binding for all floors

So far, what we have done here for the sensor indicator is associated it with only one floor. While we have gaze enabled for all floors already, and we are getting the reference for all floors while gazing on them, the only job left is to attach the indicator to each floor.

To enable that, first create a *prefab* of *icon_group* of indicator by dragging the set of Game Objects inside the `Asset` folder. Then, we can achieve the displaying part against each floor in the following two ways:

1. Place the prefab associated with each floor and show them only when gaze is entered (`OnFocusEnter()`). The data binding will take place automatically, as we already have tagged the indicator within Prefabs.
2. You can also instantiate the prefabs on the fly and display where the gaze is.

The following image shows the Sensor indicator for different floors implemented with the first approach:

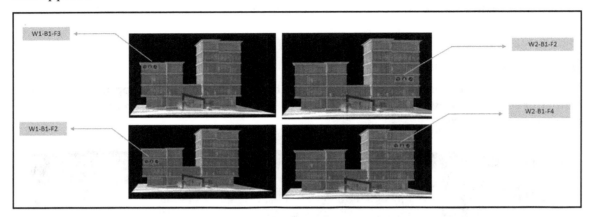

Sensor indicator for different floors

Adding actions to floors – applying gesture

Another set of requirements of this application development was to display a detail of each flat along with the indicator. In this section, we will implement the Air Tap gesture on each floor and that invokes a separate panel with some information for inspections.

Adding Air Tap on floors

To add an Air Tap gesture, we need to write custom script such as what we learned during our previous project; before adding the custom script, go to the `Assets` folder, then to **HoloToolkit** | **Input** | **Scripts** | **InputSource**, and attach the `GestureInput.cs` with `Root` object in the Object Explorer.

1. From the `Assets | Scripts` folder, navigate to the `FloorInteractionManager.cs` file and open it inside *Visual Studio*.
2. Implement another interface `InputClickHandler` for the `FloorInteractionManager` class that implements the *OnInputClicked()* methods that invoke on click input:

   ```
   public void OnInputClicked(InputEventData eventData)     {    }
   ```

At this point in time, if you run the application and place a breakpoint in the `OnInputClicked()` method, whenever you do an Air Tap in any of the buildings, you will find the breakpoint hits. Now, our job will be to display details when the air tap happens.

Displaying detailed information

In this section, we will create a small dialog panel using *Unity UI* elements, bind the data with panel elements, and finally map them with a floor gesture so that, along with sensor indicators, whenever a user taps, they can take a look at details on information, such as current temperature, floor number, or any additional details that you may plan to send via services in future.

Designing the Panel layouts

The *Dialog Panel* consists of a set of UI elements including *panel, text, and image* controls and everything is wrapped inside a canvas:

1. Start designing this by adding one **Panel** in **Object Hierarchy**, which will add a **Canvas** as a parent control
2. Then, add three sets of images for **Fire**, **Temperature**, and **Smoke,** along with a text control below to each image
3. Place a **Title** text and an **Image** for the close option

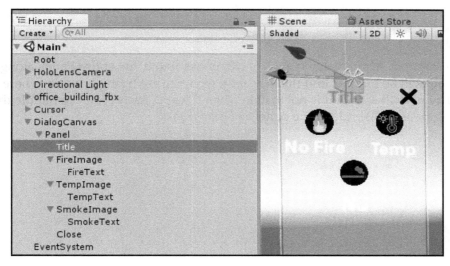

Designing the Panel Control

To apply images to image controls, you need to convert the images into Sprite. In the downloaded solution for this project, you will get the converted images into Sprite. In case you want to change the images, make sure that you convert it to Sprite by selecting the **Image Navigate to Inspector** window | and changing the **Texture** type to **Sprite** (**2D** and **UI**) from **Default**.

Once the basic design is done, tag all the text controls to **BuildingTitle, FireText, Temp Text**, and **SmokeText** for **Main Title, Fire, Temperature**, and **Smoke** elements, respectively.

Grouping UI elements – adding Canvas Group

Our Panel design is ready. We need to bind the data with the panel and need to control the visibility of all controls when we click on the close button or gaze at the other floors; this is typically how the dialog works.

Instead of controlling each individual child control, we can achieve it by adding a **Canvas Group** component in Panel. With that, the whole child UI elements can be handled together:

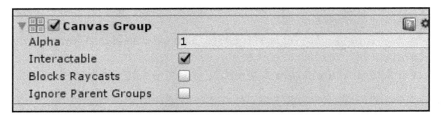

Adding Canvas Group components for Panel

If you have noticed, **Alpha** value is set to 1. If you set it to 0, the entire Panel, along with all child controls will be hidden, and that's exactly what we are trying to achieve here. Set the Alpha to 0 by default so that it is not visible when the app starts (you can handle that part using the script as well) and tag the panel as *DetailsPanel*.

Binding Data with Panel

Binding the data for panel would be very much like what we have done for sensor indicator objects. We will get this data in place during gazing at the floor, which means inside the `OnFocusEnter()` method:

1. From the `Assets | Scripts` folder, navigate to the `FloorInteractionManager.cs` file and open it inside *Visual Studio*.

2. Go back to the `OnFocus()`. Enter method and add the following code block:

```
Text buildingTitle =
    GameObject.FindWithTag("BuildingTitle").
    GetComponent<Text>();        buildingTitle.text =
    floor.FloorNumber;
```

This takes reference for BuildingTitle control and sets the values of *Building Title*.

3. Navigate to the `SetFireIndicator()` method and add the following code block:

```
Text fireMessage =
GameObject.FindWithTag("FireText").GetComponent<Text>();
    if (isFireIndicator)
    {
        ......
        fireMessage.text = "Yes";
    }
    else
    {
        ......
        fireMessage.text = "No";
    }
```

Here, we will set the values for FireText panel control based on fire indicator values coming from the sensor.

Similar to how we implemented it for **Fire**, you can do so for **Smoke** and **Temperature** as well.

Invoking the Panel

The panel is ready and data binding is also completed; the next part would be invoking it when we are tapping it on any of the floors. We have already created the method definition for `OnInputClicked()`, which is invoked when there is an Air Tap. Update the method with the following line of code:

```
public void OnInputClicked(InputEventData eventData)
  {
         GameObject detailsPanel =
              GameObject.FindWithTag("DetailsPanel");
       CanvasGroup detailsPanelRenderer =
              detailsPanel.GetComponentInChildren<CanvasGroup>();

       if (detailsPanelRenderer.alpha == 1)
          detailsPanelRenderer.alpha = 0;
       else
          detailsPanelRenderer.alpha = 1;
  }
```

When an Air tap is done, first it takes the references for the *DetialsPanel* and finds the renderer for *CanvasGroup* components. In the next step, it sets the alpha value of the renderer to 1 if it is not visible, or else to 0.

Binding the values for services data are already done as soon as you have gazed over the floor.

Closing the Panel

Implementation of Closing Panel would be a small exercise for you to consider using Gaze and Gesture; we will try to keep this short and simple here:

1. Add a box collider with the cross images inside the panel.
2. Add a new script called `DialogCloseHandler.cs` inside the `Script` folder.
3. Open the class in Visual Studio and implement the `InputClickHandler` interface.

Add the following line of code inside the `OnInputClicked()` method that just takes reference of the *DetailsPanel* Game Object and sets the alpha to its *CanvasGroup's* renderer components to 0:

```
public void OnInputClicked(InputEventData eventData)
    {
        GameObject detailsPanel =
            GameObject.FindWithTag("DetailsPanel");

        if (detailsPanel != null)
        {
            CanvasGroup detailsPanelRenderer =
                detailsPanel.GetComponentInChildren<CanvasGroup>();
            detailsPanelRenderer.alpha = 0;
        }
    }
```

Finally, attach the script with Close image control from the **Inspector** windows. That's it.

Air Tap on Lenses - see it in action

Save the scene and build and run the solution in the emulator once again. You should be able to see the cursor on each floor and its sensor data. Once you tap on any of the floors, a panel will appear and you will see the details of the values.

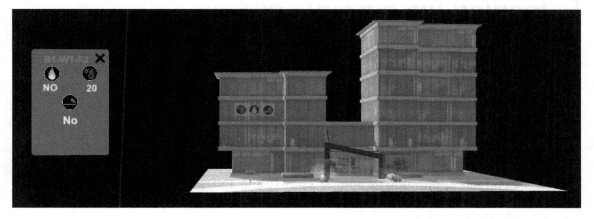

Floor sensor information with details panel

In the panel, we have changed the images as we did for the Sensor indicator. We just binded the text values. Additionally, if there is anything else than the data you are getting from service, you can just bind that too.

Deploying your app onto a device

There are several ways of deploying the application, and we discussed these in detail in `chapter 4`, *Explore HoloLens as Hologram - Developing Application and Deploying on Device*, where we developed our first project, *Explore HoloLens*:

- Deploy using Visual Studio
- Deploy using HoloToolkit Build Window
- Deploy using Device Portal
- Deploy using Unity Remoting

Use any of these, deploy your Remote Monitoring application into HoloLens, and run the application. You will find a 3D Model of a building in front of you. Gaze in to any of the floors; you should be able to visualize sensor indicators and tap on them to see the panel with their details.

 Make sure that your device is connected to the internet so that it can access the Web API

Extending the current solution

The objective of this project was to build an enterprise scenario with HoloLens with integrated IoT. We focused on the connected scenarios and building a holographic with a 3D Model of a building. With respect to the holographic application, we focused only on Gaze and Gesture; however, there are several enhancements that you can do to extend this solution--here are a few of them.

Adding Voice Command

You can extend this application by adding Voice Command. Extension scenarios could be, for example, a command to display details of a room number and your holographic app shows the details of the room.

3D Model placement as table top mode

You can extend this application by adding the capability to place the building 3D Model on any surface, such as a table top or floor. This gives a more visual experience, rather than a 3D Model hanging in the air; that doesn't provide an impactful, rich experience.

Additional interaction with services

You can plan to extend the integration of more sensor data, emergency services, building public speaker systems, and so on:

- Integrating with more sensor data, such as building entry sensor data, water supply and quality check, control, and usage of electric devices
- Integrating with **Close Circuit Television** (**CCTV**) cameras and pulling the direct video feed within the application
- Integrating with emergency services such as *Fire Department, Police department*

Summary

This chapter covered an end-to-end development aspect of a holographic application with IoT scenarios. Overall, the chapter was segregated into three different aspects of development. First, we discussed setting out a backend infrastructure where we simulate sensor data in *Azure Event Hub* and pass it through Azure Cosmos DB - SQL (DocumentDB) by parsing the raw data through Stream Analytics Jobs. Secondly, we discussed exposing a Web API for the sensor data stored into Azure Cosmos DB - SQL (DocumentDB), and finally, building the holographic application that connects with Azure Services to display live data in our application.

With all that covered, this chapter builds up a solid foundation for building a different connected scenarios application with unlimited possibilities using HoloLens and Microsoft Azure stacks. We will take this learning forward and in the next chapter, we will build another connected holographic solution.

7
Build End-to-End Retail Solution - Scenario Identification and Sketching

In the previous chapter, we successfully developed and deployed our first connected holographic application, visualized near real-time data using HoloLens. The application we have developed and tested connects with backend services, pulls data in real time, and incorporates that data feed within holographic application. However, in the last project, we only pulled the data and updated the rendering of objects based on data feed.

In this chapter and next chapter, you will learn to build another connected holographic application that will not only pull data from external system in real time and display it but will also post the data or user actions back to services. Along with that, you will also learn the scenario, where we will download assets in real time, rather than predefining them within the application. For this application development, we will follow the familiar journey of holographic application development:

1. Identifying new scenario or application requirements.
2. Detailing out that scenario through sketch and plan phase.
3. Creating 3D Assets.
4. Developing scene in Unity3D and applying scripts.
5. Connecting this application with back end service, pulling real-time data from service, and updating holograms based on data received from external service.
6. Capturing user choice and post data back to services for further action.
7. Finally, deploying it on HoloLens device and testing it with near real-time data.

This journey of connected holographic application development is again divided into two chapters, this current one and the next one. In this chapter, you will learn about identifying and prioritizing scenarios as a part of the envisioning exercise; finalizing one scenario; and doing the story boarding, sketching, and 3D Modeling, along with connected solution architecture overview.

In the next chapter, you will learn about Unity3D project, importing assets into Unity3D, scripting to implement gestures, connecting with back end/external service, pulling data from service, updating holograms based on data received from service, posting back to back end services, building complete solution, and deploying it on the device.

The following are the topic that we will cover in this chapter:

- Envisioning
- Scenario elaboration
- Sketching
- Solution overview
- 3D Model overview

In the following chapter, you will continue this same scenario and do the development and deployment on HoloLens device.

Envisioning

Let's assume you are working for XYZ Inc., which is a leading online home furnishing store that provides a huge range of stylish options for home furnishings. XYZ's retail online store is doing well, but they want to provide more personalized experience to their consumers. Right now, users search, select, and order products using customer's online web portal and mobile applications. However, their product's return rate and exchange rate is very high, and products are mostly returned or exchanged due to its mismatch with customer's current home furnishings or space restrictions. The leadership of XYZ Inc. wants to change the whole user experience, and wants the user to experience product within their home environment even before placing the order, which will result in shrinking the whole sales cycle. It will reduce the number of exchange events and ultimately reduce the cost of the delivery.

To achieve this vision of XYZ Inc., the first thing you will need to do is understand their vision better; for that, you schedule a meeting with company leadership and stake holders. In this meeting, you will get to know about their vision and requirements in detail. You can use familiar brainstorming ideas, such as sticky notes session, to discussion requirements and prioritize those ideas.

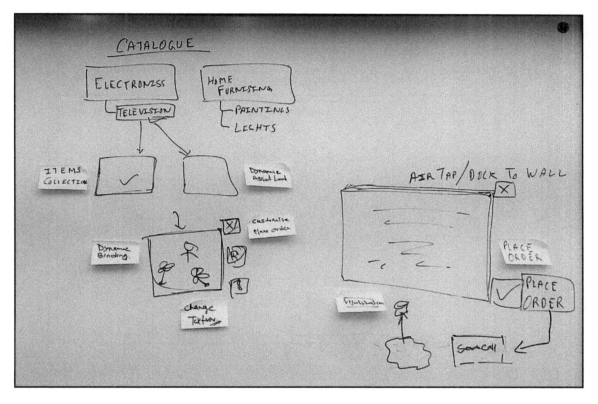

Brainstorming ideas for holographic retail experience

Scenario elaboration

From the brainstorming session with customer stakeholders, you got to know about your customer's vision toward transforming retail business, and taking products closer to consumer's home. To achieve this, customer stakeholders want to leverage MR devices, and empower consumers to visualize prospective products within their home atmosphere/environment.

Now, to achieve this vision, you need a holographic application that can connect with existing XYZ Inc.'s merchandise store, loads assets dynamically within holographic application, and let the user visualize products within their home--not only visualize but also compare, select, and place order for products from within the holographic application.

So, from here, you can break down the scenario or solution into three different categories:

- Integration with merchandising system.
- Holographic application--visualization and order placement.
- Integration with order management.

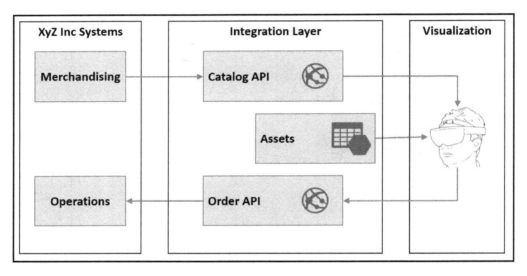

Solution overview

Catalog and merchandising

In an online retail business, catalog and merchandising are an essential part of the system that need to be kept updated at the real time and can't have any latency in the system. So, to show the latest catalog of the available products, our plan is to connect with the merchandising system of the company. The holographic application will connect with custom catalog API, and catalog API will be responsible for connecting with the merchandising system and return the latest catalog information to the holographic application.

Visualization

Within the holographic application, the user should be able to select product and place it anywhere within his real home environment. Application development will be a one-time activity, that is, it can't have assets or 3D Models for all possible products, as new products will get added within merchandising system on a daily or weekly basis. So, the application should have a mechanism to download new assets and then present it to the user. This will enhance user experience, as the user will be able to visualize new products every time.

When the application is launched, a user can view hologram of the catalog, that is, a list of categories such as, electronics, room furnishings, and much more. The user can further select any category, and subsequent items get visible for selection--for example, if the user selects electronics, the application shows a list of television set to choose from. Similarly, when a user selects home furnishings, the application shows products, such as, wall hangings, lights, paintings, and so on.

Now, if the user selects a particular item, the application should download its corresponding 3D Model from the Asset store, and then dynamically load the asset within the application. When the asset is loaded, the user should be able to place that asset anywhere within the real environment, for example, placing the television on wall of their room. The user visualizes the holographic television--if they like it, then they places an order for this television, otherwise closes this asset and opens another item from the catalog.

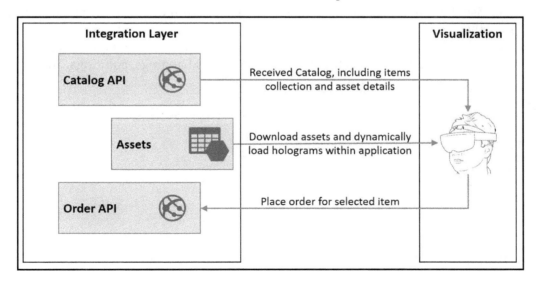

Dynamically load assets to maintain updated catalog

Order placement

After the user has visualized the product and selected the product that he wants to buy, application will have functionality to place order from within the holographic application. To enable this functionality, the user will be presented with an option to place order for the product, and to place the order, application should connect with XYZ Inc.'s operations system to place an order. If the order is placed successfully, confirmation is shown to the user.

Sketching the scenarios

Next step after elaborating scenario details is to come up with sketches for online retail using holographic application scenario. As we discussed in previous project sketching phase, there is twofold purpose for sketching, first it will be input to the next phase of asset development for the 3D Artist, as well as it will help in validating requirements from the customer, so that there are no surprises at the time of delivery. For sketching, either the designer can take it up on its own and build sketches or can take help from the 3D Artist.

Let us start with the sketch for the primary view of the scenario, where application starts and user views the retail catalog.

- Visualize and select primary categories like electronics, home furnishings, and others
- Select primary category, and visualize subs categories. Like if user select home furnishings, then sub-categories like lights, wall hangings, paintings, and so on

Sketch for user viewing hologram for catalog

Sketching - interaction for catalog and items

While viewing the catalog, user select one sub-category like wall-lights, then show him list of all lights available in the stock. User can review this list of wall-lights, and either closes the list and move to another category within catalog, or select one of the light and let the application download this asset.

The user can air-tap on any of the item within the list, and the application will start downloading latest asset for that item. After the asset is downloaded, the corresponding hologram will be available to the user, and user will have to place it at the desired location.

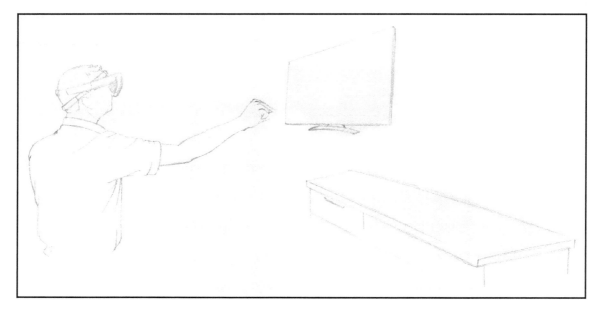

Selected asset and place it at desired location

Sketching - interaction with 3D Model

After the asset is downloaded, application shows the 3D Model. A user can Air Tap and place it within the room, as per his liking. After that user, can visualize that item from different angles.

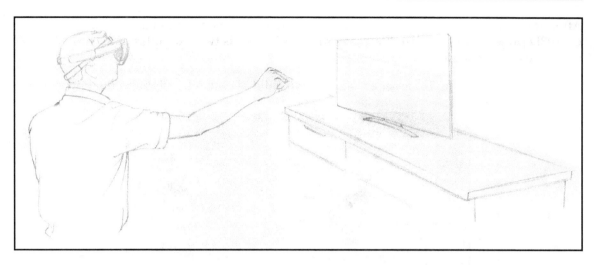

Visualize 3D Model after placement

3D Model

In previous section, you have identified scenarios and detailed them out through the sketching process. Next step is to develop the 3D Asset for catalog and retail items, that you are going to use in the next chapter for developing retail holographic application.

While developing previous project, we have already gone through detailed process of developing 3D Models. So, you have two options, they are as follows:

- **Create new 3D Model**: Apply 3D Model creation process you learned while developing previous project and develop new 3D Model for catalog and assets.

Or

- **Download pre-build 3D**: Asset using following link – `https://github.com/Pac ktPublishing/HoloLens-Blueprints`.

After you have downloaded these assets, you can import them within assets collection of Unity3D project and visualize their structure. Following is two examples of 3D models that you have downloaded.

- **Wall Lamp:** This asset consists of two child 3D Models – light holder and lamp

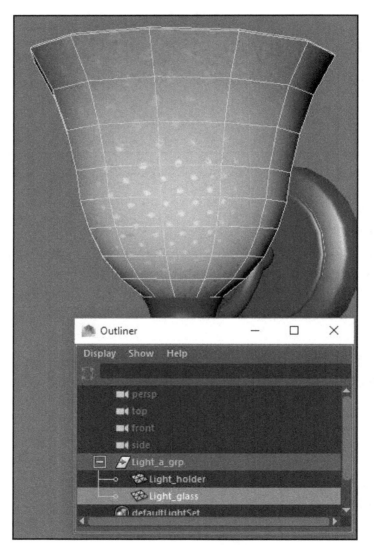

3D Model for the Wall lamp

- **Wall Painting:** This asset is again divided into two child 3D Models, specifically painting frame and painting inside the frame

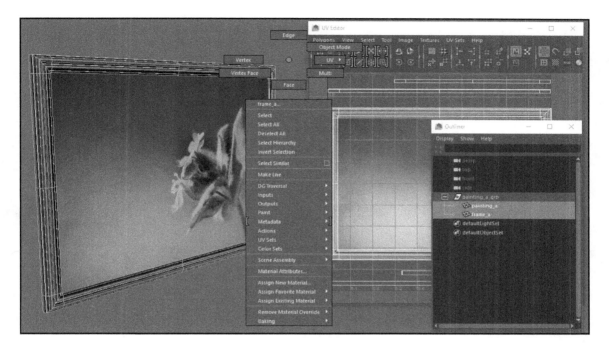

3D Model for the Wall Painting

Summary

In this chapter, you started the development process of creating first end to end connected retail solution, and learned development aspects like envisioning, scenario elaboration, sketching, architecture and solutioning for online retail solution.

In the next chapter, you are going to learn about creating Unity3D project, importing assets into Unity3D, scripting to implement gestures, building complete solutions, connecting with back end services for merchandising, and placing orders.

8
Build End-to-End Retail Scenario - Developing Application and Deploying on Device

In this chapter, we will continue our journey toward building the holographic application for the retail solution, and we will start from where we left in the previous chapter. In the previous chapter, we begin by envisioning this retail solution; we discussed different types of scenarios that could fit in a *Mixed Reality Holographic Model*, identified few of the scenarios to be developed, finalized the scenario, and did scenario sketching. In the envisioning phase, we discussed requirements and scenario identification for the application, and talked about several scenarios, such as downloading different products on demand, placing them in a home, visualizing them, and finally placing an order for the selected product. As a part of the scenario prioritization phase, we listed the scenarios to be taken towards development phase. We took a step further towards sketching to visualize and revalidate the identified scenarios, which illustrate a kind of vision demonstration of the solution.

Now, we will take a step forward towards the development of the application with some of the identified scenarios. This chapter will cover all the steps, starting from making reusable 3D Assets, setting up of the backend solutions with Azure, building up the services, developing the holographic project with Unity3D and finally connecting it with Azure Services till deploying it onto the device. A brief overview of the different aspects of this chapter are as follows:

- Preparing 3D Assets for dynamic loading
- Building backend Azure Solutions
- Building a holographic application
- Connecting a holographic app with Azure
- Downloading assets and rendering them as holograms on the fly
- Configuring, building, and testing connected a holographic app
- Extending the application

By end of this chapter, you will cover the complete end-to-end development process that required building connected solution for a retail holographic application.

After completing this solution development, you will learn how to download and render your hologram as and when needed, perform a bidirectional communication with a holographic app and backend services. With this knowledge, you can build an application that is lightweight, dynamic in nature, and scalable in different aspects of several enterprise customer scenarios.

This project required several assets, including 3D Models for product, main menu, toolbar controls, and so on. Before you proceed further toward development, you must download all the project related assets from `https://github.com/PacktPublishing/HoloLens-Blueprints`.

Getting the 3D Assets ready

A retail solution may have several products. The 3D Assets which we are referring to here, are referring to those assets which we are going to see as a product hologram in the holographic application. At a very first steps of the solution development you need to make those assets ready and make them shareable, so that you can download them on-demand and use them within the application. Once assets are ready, we upload them on storage and get them downloaded as and when they are needed by the application.

Here, we will be generating the Asset Bundles for on demand loading of assets from the storage.

Unity Project and Asset Bundle setup

We will take care of the process of generating Assets Bundles as part of different a Unity Project altogether. Create a new Unity3D project perform the following steps:

1. Navigate to `Assets` folder in the **Project Explorer** window.
2. Create a folder named `AssetBundles` inside `Assets` folder to store all the created bundles.
3. Add a new `C# scripts` from `Assets` | **Create** | `C# Script`, and name the script as `createAssetBundles.cs`.
4. Add the following line of code in the script file.

This is very common and standard code for generating Asset Bundles in Unity3D. What the following code does is, it adds a menu item -Build Asset Bundles under the `Assets` menu and when you select the menu items it generates the Asset Bundles for the defined assets which we will cover in the next section.

```
#if UNITY_EDITOR
using UnityEditor;
#endif
public class CreateAssetBundles
{
#if UNITY_EDITOR
    [
    MenuItem("Assets/Build AssetBundles")]
    static void BuildAllAssetBundles()
    {
        BuildPipeline.BuildAssetBundles("Assets/AssetBundles",
         BuildAssetBundleOptions.None, BuildTarget.WSAPlayer);
    }
#endif
}
```

You must make sure the **BuildTarget** is set to **WSAPlayer**. Because we need to generate the assets bundles based on the targeted platform. All holographic applications are **Universal Windows Application** (**UWP**), the type of build target has to be set as WSAPlayer (Build an Windows Store Apps player) . While WSAPlayer BuildTarget can be used for any Windows Store Apps, in this case we are specifically targeting UWP Apps running on Windows 10 Platforms.

Importing 3D retail assets

For the development of this solution, we will be using following three assets which you can download as `FBX` file from the project download folder.

- TV
- Light
- Wall paint

Import all of them in your Unity project using **Import Assets** options. Once all these assets are imported into the project, add them into your scene. For all these assets, set the transform position to 0,0,0, so that, when we render them in our holographic app, it starts on the same coordinate.

Assets Bundles generation

Once the 3D Assets have been added in the scenes, follow these steps.

1. Select the asset to assign it to a bundle, for an example select the **TV Model**.
2. From the **Inspector** window scroll down to the bottom, you should see a section to assign an asset to a bundle.
3. Provide the name of the Bundle, by selecting the drop down and adding new name. In this case, we have used asset bundle name **tv**.

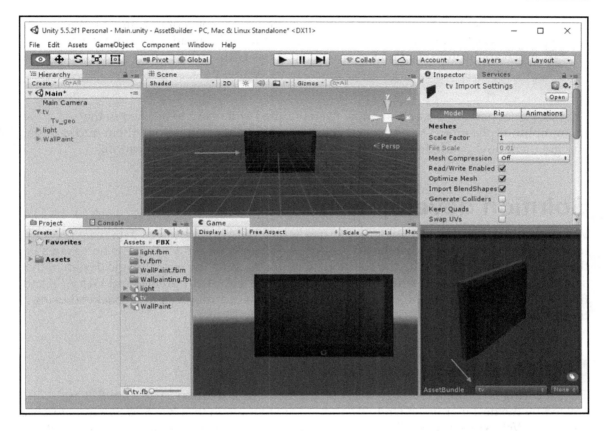

Assign asset to an Asset Bundle

Like what we have just done for TV, select **Light and Wall paint** one by one and provide the asset bundle name.

- Once done, from main menu, navigate to Assets | Build Assets Bundle to generate the Asset Bundles for all the assets.

Unity will show a pop up message during the process of Asset Bundle generation, and once done you can navigate to Assets | Asset Bundles folder, where you can see all the generated assets bundles. Note that we can bundle multiple assets together as well. In this case, we have keep them as separated for simplicity purposes.

Asset Bundles Generation is a very important concept in Unity3D. With Asset Bundles, you can dynamically load and unload assets into your application. To know more, Refer to this URL: `https://unity3d.com/lea rn/tutorials/topics/scripting/assetbundles-and-assetbundle-man ager`.

That's all from here. Our assets are now been bundled and ready to be used in the holographic application. Once backend solution is ready, we will upload them into storage and get them downloaded as and when needed in the holographic application.

Solution development

We are building a connected holographic app, which sends and receives data from services hosted on the cloud, download the 3D Assets on demand and render them as holograms on the fly. Let's focus on building the solution by understanding the overall logic design of the solution. After that we will setup the backend solution with *Azure Storage* and *Web API*, and finally we will develop the holographic application that communicate with backend using the developed API's.

Logical design diagram

The following is the logical design diagram for the overall solution. *Assets Bundles* are created by the designer using Unity3D and uploaded them in to *Azure Blob Storage* reference URL. The Web API exposes API to get data from Azure Cosmos DB--SQL (DocumentDB), which has the reference URL of the Assets along with other product details. Holographic apps access the API to get product details, along with the product asset URL. Holographic app downloads the assets from Blob storage directly.

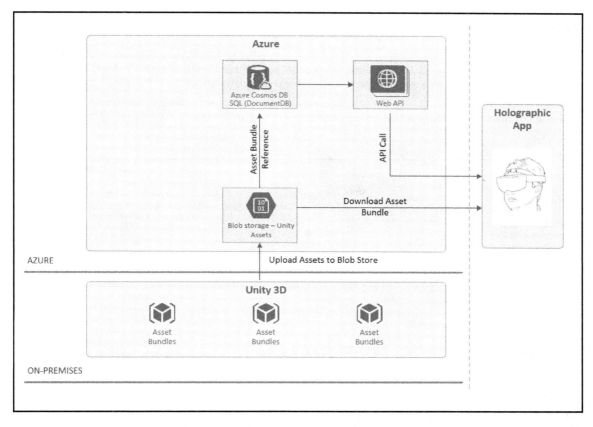

Logical design diagram for the holographic retail solution

Building the backend Azure Solution

To build the backend infrastructure, as per our design, we need to create an Azure Blob Storage to store all Asset Bundles, and then create a Azure Cosmos DB--SQL (DocumentDB) database for product details along with the Product URL from Azure Blob Storage. Finally, we will create a Web API application, that will read data from Cosmos DB and will make it available to the holographic application for receiving the data and send a call back to services.

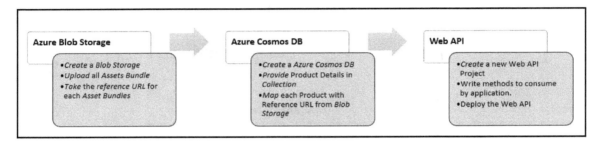

Task for backend solution setup

Creating the Azure Blob Storage

Log in to your **Azure Subscription**, and create an Azure Storage using **Create Storage Account** and by selecting all necessary parameters, like *name, resource group,* and so on. There are a total of five types of Azure storage - *Blobs, Files, Disks, Table and Queues*. In our case, we will be using *Blob Storage. Blob Storage* is used for store documents, pictures, videos, and other unstructured large text content or binary data.

Once the storage account is created, follow these steps:

1. Select the **Blobs** storage services option.
2. Select the add **Container** options.

3. Provide the container **Name** and set **Access type** to **Blob** and click on **OK**. This will create the **Blob Container**.

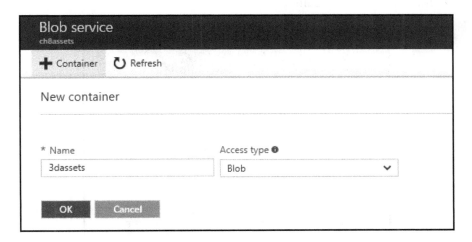

Create new Blob container

4. Once Blob container is created, navigate to the container and upload all the assets bundles that we have created in the **Asset Bundle Generation** section.

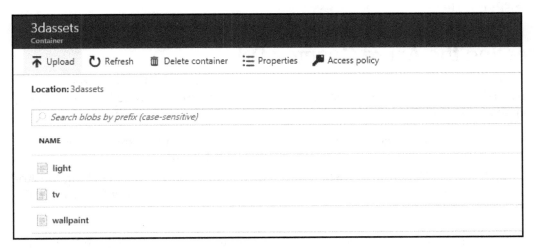

Upload Assets Bundles to Blob Storage

5. Select each of the uploaded item in the list, and from the **properties** window get the URL of the blob storage. We will use this as product reference in the Azure Cosmos DB--SQL (DocumentDB).

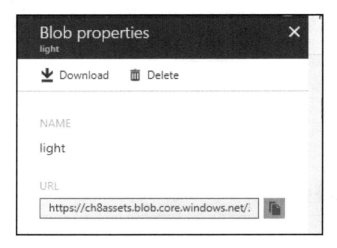

Referring the Blob storage URL

With that, we are done with configuring the Blob Storage and uploading our assets. In the next step, we will create the Azure Cosmos DB.

Creating the Azure Cosmos DB

The next step to set up our backend infrastructure is to create the Cosmos DB--"*SQL DocumentDB*" database with product details and refer the Blob Storage URL for respective assets; to do that follow these steps:

1. From **New**, select the **Azure Cosmos DB** services and create new DB by providing name, resource, and other required details.
2. Once the **Cosmos DB** instance is created, add a collection by clicking the **Add Collection** button and add a database with in it by providing following *JSON Content* with `Product Name`, `URL` and the `AssetName`.

Replace all the AssetURL in the following JSON with the Blob URL you get in previous section.

```
{
```

```
"Id": "ProductCatalog",
"Products": [
    {
        "Name": "TV",
        "AssetURL":"https://ch8assets.blob.core.
                     windows.net/3dassets/tv",
        "AssetName": "TV"
    },
    {
        "Name": "Lights",
        "AssetURL":
                 "https://ch8assets.blob.core.windows.net
                     /3dassets/light",
        "AssetName": "light"
    },
    {
        "Name": "WallPaint",
        "AssetURL": "https://ch8assets.blob.core.
                     windows.net/3dassets/wallpaint",
        "AssetName": "painting_a"
    }
],
"id": "69c79518-fd7f-803a-e931-c007d6b1deb9"
}
```

 Remember, the `AssetName` must be the same as the original asset name from which you created the Asset Bundles.

3. Note down **URI**, **PRIMARY KEY**, *Collection name* and *Database name* for further reference, which you can get from **Settings| Keys.**

Taking Reference of URI and Primary Key for Azure Cosmos DB

With that, Cosmos DB part is over. Let's move the creating the Web API Project task.

Developing the Web API

The *Web API* consumes the *Cosmos DB* data that we just created and exposes the *REST API* *be* consumed by other applications. Here are the steps to be followed to build the service.

1. Open Visual Studio and create web application, by using context menu **File| New Project| Web | ASP.NET Web Application (.NET Framework)**.

2. From the **New ASP.NET Web Application** dialog window, select **Azure API App** option.

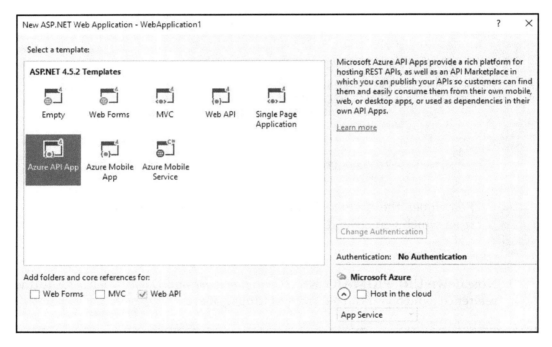

Azure API App selection

3. Add Data Models for *Product* and *Product List* as follows by creating two different Class `Product.cs` and `ProductList.cs`.

```
public class Product
    {
            public string Name { get; set; }
            public Uri AssetURL { get; set; }
            public string Price { get; set; }
            public string AssetName { get; set; }
    }
```

```
public class ProductList
    {

        public string Id { get; set; }
        public List<Product> Products { get; set; }

    }
```

4. Within **Project**, add a new class, say `ProductListCreator.cs` , which is responsible for connecting with SQL-DocumentDB and reading product details

5. Within `ProductListCreator` class, add following constant strings for connecting with SQL-DocumentDB.

```
private const string EndpointUri =
"https://[endpointname].documents.azure.com:443/";
private const string PrimaryKey = "[primary key]";
private const string DatabaseName = "[database name]";
private const string CollectionName = "[collection name]";
```

6. Now within `ProductListCreator` class, add a new method for fetching data from SQL-DocumentDB.

```
public ProductList GetProductList(string id)
    {
        this.client = new DocumentClient(new
            Uri(EndpointUri), PrimaryKey);
        FeedOptions queryOptions = new FeedOptions {
            MaxItemCount = -1 };
        IQueryable<ProductList> prodQuery = this.client.
            CreateDocumentQuery<ProductList>(

    UriFactory.CreateDocument
        CollectionUri(DatabaseName,
        CollectionName), queryOptions)
        .Where(f => f.Id == id);
            for each (ProductList product in
              prodQuery)
            {
                    return product;
            }
            return null;
    }
```

7. Add a new controller by right clicking on the `Controllers` folder within the newly created project, select **Add | Controller** from the context menu, and name it `AssetsController`.

8. Within this controller class, add following method which returns the `ProductList` when `GetAsset()` is invoked.

```
public IHttpActionResult GetAssets()
        {
                ProductList planlist = new
ProductListCreator().LoadPlanList("ProductCatalog");
        if (planlist == null)
                {
                        return NotFound();
                }
                return (Ok(planlist));
        }
```

9. Add another controller, by right clicking on the `Controllers` folder within the newly created project, select **Add | Controller** from the **context** menu, and name it `PurchaseController`.

10. Add the following `Post` method with in the Purchase Controller.

```
[HttpPost]
public IHttpActionResult AddToCart()
    {
            return Ok("Added To Cart");
    }
```

11. Build the solution and deploy it to Azure.

Once your service is hosted and deployed on Azure (We have dicussed the deployment process of Azure Services in the `Chapter 6`, *Remote Monitoring of Smart Building(s) Using HoloLens - Developing Application and Deploying on Device*) , you can hit the end points (like - `http://<servicename>>.azurewebsites.net/api/assets`) with any of the *Rest API Client Tools*,you should be able to view the JSON data in a format like the following:

```
"Products": [
    {
        "Name": "TV",
        "AssetURL":
            "https://ch8assets.blob.core.windows.net/3dassets/tv",
        "Price": null,
        "AssetName": "TV"
    },
```

Once the service is deployed and end points return the preceding JSON data, our backend solution is ready. Now we will move to the next steps of the solution development we will start building up the *Holographic Solution using Unity3D* and then integrate with our backend services set up so far.

Setting up the holographic project

Setting up the Holographic Project--includes creating a new Unity3D Project, setting up Unity editor, and using the HoloToolkit. In short, these initial setups are similar for all Holographic applications and, over past few chapters, you have already experienced it.

 So far, we have developed two end-to-end applications and by now you are familiar with all these steps. In this section, we will mostly cover it at a very high level. Anytime you need to check out details, you can refer to the Chapter 4, *Explore HoloLens as hologram - Developing Application and Deploying on Device* or Chapter 6, *Remote Monitoring of Smart Building(s) using HoloLens - Developing Application and Deploying on Device*. Also, Holographic Interaction Models like Gaze and Air Tap would be cover at a high level while implementing this solution, as we already covered them during previous exercise.

At a high level, perform the following steps:

1. Start a new instance of **Unity3D**, and **Create** a new **Unity3D** Project. Let's name it as HoloRetail.
2. One the new project is created, **Import** the **HoloToolkitUnityPackage** Package.
3. Apply the **Project Settings** for **Holographic Application** by setting the **Build Settings**, and **Enabling Virtual Reality** option for of the application.
4. Update the initial scene by removing the existing *Main Camera*, then navigate to Assets | **HoloToolkit** | **Input** | **Prefabs** and add the HoloLensCamera into the Scene.
5. Finally, **Save** the scene.

With this, your initial Unity3D holographic app is ready and you can start building top of it.

Importing the main menu 3D Model

Once the initial Unity3D solution is ready, the very first thing we will do is importing the main menu. You already have that menu object model inside the `download` folder named as `Menu.fbx`.

1. In the **Project Explorer** windows, right-click | **Import New Assets**, and import the `.fbx`.
2. Once import is completed, drag and drop the menu into **Object Explorer**.
3. Set the *Transform Position (x,y,z)* to `(0,0,10)`.

Main menu 3D Model for HoloRetail application

Exploring the main menu

If you expand the **menu** Game Object, in **Object Hierarchy**, you will find it has four different components. One of them is for title **HoloRetail**, and other three Game Objects represent individual product items.

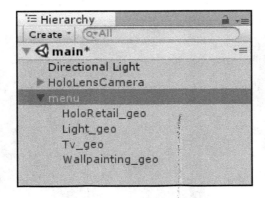

Main menu Game Object in Object Hierarchy

Select the individual menu items from the Object Hierarchy and renamed them to **Lights**, **TV** and **WallPaint** accordingly. This is just for easily identifying the menu items by object name when users tap on it.

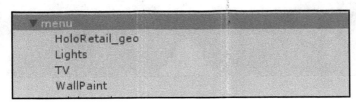

Menu renamed objects

You can alternatively do this by *tagging* them with different name. However, in that case, the script would change to get the object by tag name.

Our next objective will be make the menu interactive and when user perform an Air Tap on menu item, it should get the respective assets downloaded from Azure Blob Storage and render them as hologram with in the holographic application.

Building , running and testing the initial solution

With that our initial solution is ready with main menu. Save the project, build it in Unity, and run the project in *HoloLens Emulator*.

Here is the very first look of your **HoloRetail** app with a main menu:

FHoloRetail app first look

 You may refer to Chapter 4 , *Explore HoloLens as Hologram - Developing Application and Deploying on Device* or Chapter 6, *Remote Monitoring of Smart Building(s) using HoloLens - Developing Application and Deploying on Device* in case you need more details on how to do the build and run the initial solution.

Making the main menu Intractable

Once the basic app is up and running, our next action is to make each menu item interactive by adding *Gaze* and *enabling Air Tap*.

Enabling Gaze with cursor

To enable the gaze for each of the menu item, follow these steps:

1. Select each of the item Game Objects (*Tv, Lights, Wallpaint*) individually and, from the **Inspector** window, add a Box Collider component to them.

Box Collider added to individual menu item object

2. Add an empty Game Object in **Object Hierarchy**. Rename it as **Root**.
3. Add the following scripts from `HoloToolkit` folder into to **Root** Object " *Gaze Manager, Gaze Stabilizer, Gestures Input and Input Manager*".
4. Navigate to **HoloToolkit | Input | Prefabs | Cursor**, and drag drop the *Cursor* prefab into **Object Hierarchy**.

With these steps performed, if you build and run the application once again, you will see the cursor whenever you gaze on any of the menu items.

To highlight the Gazed items along with the cursor indicator, do the following:

1. Create a folder called `Scripts` inside `Assets` folder.
2. Add a new script by navigating from context menu | **Create | C# Scripts**, name it `TapHandler`.
3. Open the script file in Visual Studio.

By default, the `TapHandler` class is inherited from `MonoBehaviour`. `MonoBehaviour` is the base class from which every Unity script derives. In the next step, Implement `IFocusable` interface in the `TapHandler` class. We have already seen this interface implement the methods `OnFocusEnter()` and `OnFocusExit()`, methods which invoke when *Gaze Enter* and *Gaze Exit* happens respectively.

Update the newly created class with following piece of code.

```
public class TapHandler : MonoBehaviour, IFocusable
{
     Color startcolor;
     Renderer rendererObject;
     public void OnFocusEnter()
     {
          rendererObject= this.gameObject.GetComponent<Renderer>();
          if (rendererObject!= null)
          {
               startcolor = rendererObject.material.color;
               rendererObject.material.color = Color.blue;
          }
     }
public void OnFocusExit()
     {
          if (rendererObject != null)
          {
               rendererObject.material.color = startcolor;
          }
     }
```

Attaching the script with menu items

Individually, attach the `TapHandler.cs` script to all the menu item object as components. With this, whenever you gaze on an item, along with the cursor, the selection of the item color would also change.

Enabling Air Tap with menu items

Once we have the cursor on the gazed item, we can enable the gesture by adding custom scripts with menu items.

1. From the `Assets | Scripts` folder, Navigate to `TapHandler.cs` file and open it inside Visual Studio.
2. Implement another interface `InputClickHandler` for the `TapHandler` class, that implement the methods `OnInputClicked()` that invoke on click input.

```
public void OnInputClicked(InputEventData eventData)
{
}
```

At this point of time if you run the application and place a breakpoint in `OnInputClicked()` method, whenever you do an Air Tap in any of the menu items--let's say for TV or Lights, the breakpoint would hit.

And that's all from here, our application main menu is now interactive and has gazed enabled with cursor and color indicator. We also implemented the Air Tap skeleton. The next job would be connecting with backend services and downloading the required assets when there is an Air Tap on any of these menu item.

API connector - connecting with Web API

In this section, we will connect our holographic application with the Web API hosted on Azure.

Defining the application data model

The data model will hold the data returns by the *Asset Rest Services*, hence the data model structure would same as what we used during development of Web API. You can use any of the *Rest Client Application* and hit the service end to get the JSON response from services and based on that you can create the data model. Follow these steps here:

- In the Unity3D Project, add a new folder inside the `Scripts` folder called `Models`.
- Add two different script file `Product.cs` and `ProductList.cs`.

Define all the classes as follows:

```
[Serializable]
public class Product
{
        public string Name;
        public string AssetURL;
        public string Price;
        public string AssetName;
}

[Serializable]
public class ProductList
{
        public string Id;
        public Product[] Products;
}
```

Make sure you add `Serializable` attributes for each class, so that it can be serialized.

 You must remember, in Unity, we must use a data field rather than using properties for any given data models where data serialization is required. Unity cannot serialize properties; that's why all the members in the data model define as fields rather than properties. Refer to `http://docs.unity 3d.com/ScriptReference/SerializeField.html` for more information related with Unity Serialization.

Enabling the IntenetClient capabilities

Enabling access to *InternetClient*, we allow our holographic application to interact with external web requests.

From the Unity main menu, **HoloToolkit | Configure | Apply HoloLens Capability Settings** and checked the **InternetClient** option to enable.

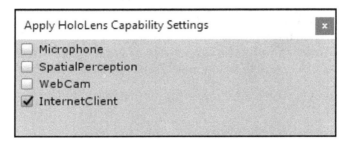

Applying HoloLens capability settings

 You can also enable this setting by navigating from **Edit | Project Settings | Players** and Under Settings for **Windows Store | Publishing Settings | Capabilities**, you will find option **InternetClient**.

API Connector

Now, we will create the connector like the *Azure Bridge* that we created during previous chapter exercise, which fetches the data from the services and map it with the *Product Data Model*, we just defined.

1. In **Unity Project Explorer** window, Navigate to `Assets | Scripts` folder.
2. Add a new script by navigating from **context menu | Create | C# Scripts**, name it `APIConnector`.
3. Open the script file in Visual Studio.

At a very first step, we will make this class as `Singleton` to use only one instance of it. You can achieve this by just inheriting the class from `Singleton` class, by passing the same class type as parameter.

```
public class APIConnector : Singleton<APIConnector>{
        void Start () {
        }
        void Update () {
        }
}
```

Getting the data

Connecting with Web API and fetching the data to our application can be done in several ways. Like other projects we have done in previous chapters, we will `UnityWebRequest` to communicating with our backend services.

> UnityWebRequest allows Unity games to interact with backend services over HTTP Request and Response.

Once data gets downloaded by the `DownloadHandler`, we need to convert it from the JSON response to Product. The `JsonUtility` helps in that conversion.

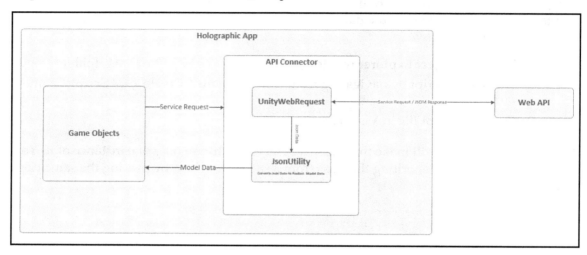

API Connector with holographic app

Here is the `APIConnector` class implemented with `UnityWebRequest`.

```
string AssetAPIURL =
    "http://<<servicename>>.azurewebsites.net/api/assets";
public ProductList products;
void Start()
{
        StartCoroutine(GetProductDetails());
}
private IEnumerator GetProductDetails()
{
        UnityWebRequest webRequest =
            UnityWebRequest.Get(AssetAPIURL);
        yield return webRequest.Send();
        if (webRequest.isError)
        {
                Debug.Log(webRequest.error);
        }
        else
        {
                string productData = System.Text.Encoding.UTF8.
                GetString(webRequest.downloadHandler.data);
                products = JsonUtility.FromJson<ProductList>
                        (productData);
        }
}
```

The `GetProductDetails()` method sends a `WebRequest` to the service endpoint and, once data returns from the service, the `JsonUtility.FromJson()` method converts the raw *JSON* data into our desired product model objects.

Attaching the API connector

Attach this script with **Root** object of our holographic app, by just dragging the script into the **Root** object, either in the **Object Hierarchy** or in the **Inspector** window.

With this, our *API Connector* is ready to connect and fetch the live data. If you run the application by placing a breakpoint inside `GetProductDetails()` method, you should be able to explore the data retrieved from the services in JSON format.

Getting holograms on the fly

This part is one of the interesting areas, where we are going to get the service returned data based on the selected product in the menu and, thereafter, we will download the right assets from Azure Blob, and finally, render it as hologram.

Handling the Air Tap on menu item

A hologram will download when we tap on the respective menu item. So, we have to write some action on the Air Tap of the menu items. Follow these steps:

1. In the Unity3D Project, from **Project Explorer** window, Navigate to `Assets | Scripts` folder.
2. Select the `TapHandler.cs` and open it using Visual Studio.
3. At this point in time,the `OnInputClicked()` method does not contain any piece of code. Update the method with following lines of code:

```
public void OnInputClicked(InputEventData eventData)
    {
ProductList prodList = APIConnector.Instance.products;
            if (prodList != null)
            {
                Product prod =
                  prodList.Products.FirstOrDefault(item =>
                  item.Name == this.gameObject.name);
                string url = prod.AssetURL;
                WWW www = new WWW(url);
                StartCoroutine(DownloadAndProcessAssets
```

```
                                              (www, prod.AssetName));
                    }
            }
```

What the preceding code block does is, when a specific menu item is tapped, it accesses the instance of APIConnector. The instance has reference of Products from the start of the application. this.gameObject.name returns the name of the Game Object which was tapped. Then, it filters the specific product from list of products. It then starts a coroutine to download the Asset Bundle.

 If you recall, when we designed the menu, we kept the menu item name the same as the respective product name and now, you must have realized how we are using them here to filter the particular product.

Downloading assets and rendering holograms on the fly

Add another method called DownloadAndProcessAssets() with following code block. It downloads the AssetBundle from the Blob URL and, once thee download is complete, we finally instantiate a new object with the downloaded asset.

```
public IEnumerator DownloadAndProcessAssets(WWW www, string
      assetName)
   {
        yield return www;
        bundle = www.assetBundle;
      if (www.error == null && bundle != null)
      {
              launchedObject = Instantiate((GameObject)bundle.
                LoadAsset(assetName));
              launchedObject.transform.position = new
                Vector3(launchedObject.transform.position.x,
                this.gameObject.transform.position.y - .25f,
                this.gameObject.transform.position.z - 1.5f);
      }
      else
      {
              Debug.Log(www.error);
      }
   }
```

With this, if you run the application again, and perform an Air Tap on any of the menu item, the same object is rendered as hologram and you can experience them as a 3D hologram like another static hologram object.

 The WWW Unity class helps retrieve contents from given URL, either in format of byte array, texture, text or even as assetBundle. `WWW.assetBundle` streams the AssetBundle contains any types of assets. In our case, it downloads the product assets from Azure Blob Storage.

 The *isDone* property of WWW class indicate if download is completed and *Progress* property returns how far the download is progressed. You can use these two properties and show an indicator during the download process of assets. The assets we used in this project are light weight, which we can be downloaded very fast. However, for larger objects, you must use a Visual Indicator.

Once we have our dynamic holograms rendered, our next job is make them intractable. Keep in mind, we uploaded the assets bundles on the Blob storage directly, there were no script associated with them. This will be another interesting opportunity to learn how we can apply script on those dynamically generated holograms and how to play around with them.

In the next section, we will attach such dynamic scripts and a toolbar will be associated with each of the product holograms

Associating toolbars with holograms

Enabling tooling aspect with holograms is another important aspect, where depending on types of holograms you rendered, you can enable different types of option buttons. In this case, for each of the hologram *TV, Light and Wall Paint* we will have a toolbar with *Remove* and *Add to Cart* icons. We will also perform some additional operations for *Wall Paint*, by adding additional button Refresh Which will change the painting within the downloaded Wall Paint Asset.

The toolbar controls

You will get this toolbar control as part of downloaded package. Follow these steps to bring it into your newly created holographic project

- In the **Project Explorer** windows, navigate to the `Asset` folder and create a new folder called `Resources`.
- Inside the `Resources` folder, Right-click | **Import Package** | **Custom Package** and then select `Producttool.unitypackage`.
- Wait till the import process is completed.

This is how the toolbox controls look. If you explore this in **Object Hierarchy**, it has three different Game Objects: **Close**, **Refresh**, and **Add To Cart**. All object items have Box Collider added, so that they can be gazed.

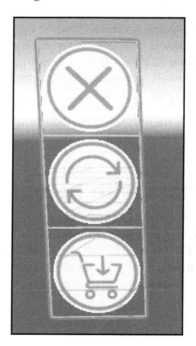

Product toolbar

 You don't need to drag and drop this control to Object Hierarchy. Yes! That's right. We are going to instantiate this toolbar on the fly by attaching them with dynamically rendered products holograms.

Instantiating toolbars on the fly

We need to attach the toolbar with each product which is getting downloaded and rendered when user Air Tap on the menu item. Navigate to `TapHandler.cs` class and add the following line of code within the `DownloadAndProcessAssets()` method post `launchedObject` is instantiated and positioned.

```
GameObject toolBar = Instantiate(Resources.Load("ProductTools",
typeof(GameObject)) as GameObject, new
Vector3(launchedObject.transform.position.x + 0.5f,
launchedObject.transform.position.y, launchedObject.transform.position.z),
Quaternion.identity);
```

In this code block, we instantiate the Toolbar by loading it from the `Resources` folder and position it next to the `launchedObject`. So that when a product, let say TV, is downloaded and rendered as Hologram you will find TV along with this control next to it.

TV with toolbar button generated on the fly

Showing controls based on types

We already discussed changing the toolbar button based on the type of product we select. We have three different buttons in the toolbar, and we want the **Refresh** button to be present only when user downloads a **WallPaint**.

Add the next set of code block just after the instantiating your toolbar.

```
if (launchedObject.name != "painting_a(Clone)")
    {
        toolBar.transform.FindChild("Referesh").
        gameObject.SetActive(false);
    }
```

With this preceding code added, the **Refresh** button will only be visible when we have a WallPaint downloaded, for other products we will have only two options: **Close** and **Add to Cart**.

WallPaint with additional Refresh Icon in the toolbar

Dynamic holograms in action

At this point of time, if you run the application once again, you should be able to check out how this hologram is getting rendered and how the toolbar is getting associated with each product. Follow these steps:

1. Build and run the holographic app in the emulator.
2. Once the app starts, you should be able to see the main menu.
3. Gaze on the menu item and do an Air Tap on any of the products; you can do the same for multiple products as well.
4. You will find products are getting rendered along with the toolbar.

 If you download different products, let's say WallPaint and TV, you will find different sets of toolbar for two different products.

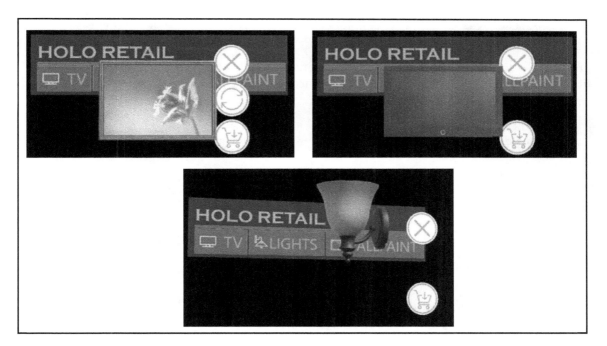

Dynamic holograms on action

Adding action to the toolbar controls

We have toolbar coming up with products and you can gaze on them using the existing cursor. However, there is no action attached to it. Now we need to add the respective script to each of these controls.

Remove product

Remove product unloads the holograms from the scene. When you do an Air Tap on the `Close` icon, it removes the associated products.

The following code block finds the `Close` Game Object from the Toolbar control and attaches an `ProductRemover` script as a component with the Game Object. Once the component is added, pass the current `launchedObject` as parameter to the `ProductRemover` Class.

```
GameObject removeProduct = toolBar.transform.FindChild("Close").gameObject;
ProductRemover remover = removeProduct.AddComponent<ProductRemover>();
remover.toberemovedObject = launchedObject;
```

You need add the `ProductRemover.cs` file inside the `Scripts` folder. The class must inherit from the `InputClickHandler` Interface and implement the `OnInputClicked()` method as shown in the following code:

```
public GameObject toberemovedObject;
public void OnInputClicked(InputEventData eventData)
  {
        try
        {
          Destroy(toberemovedObject);
              Transform gg =
               transform.parent.gameObject.transform.parent;
              Destroy(transform.parent.gameObject);
}
        catch (Exception ex)
        {
              Debug.Log(ex.Message);
        }
    }
```

When there is an Air Tap on the **Close** button, it takes the reference of that specific object as `toberemovedObject`, and destroys the object along with its own parent object - which is nothing but instance of the toolbar itself. When you tap on the **close** button, it closes the product instance.

The `Destroy()` method removes a Game Object, component, or asset which is specified. Refer this URL `https://docs.unity3d.com/ScriptRe ference/Object.Destroy.html` to learn more about `Destroy()` method.

Add to Cart

The **Add to Cart** button click sends an call back to our Web API to indicate you are adding the product to your cart. While in the backend we are not performing any specific action to add this to the cart, here we will see how we can send a call back to services as POST request.

Add the following line of code just after the previous code block.

```
GameObject cart = toolBar.transform.FindChild("AddToCart").gameObject;
ProductAddToCart prr =
        cart.AddComponent<ProductAddToCart>();
prr.addToCartProduct = this.launchedObject.name;
```

The preceding code does the same operation in the toolbar control as we did for the `Close` button. Here, we find the child control name `AddToCart` and attached a Script called `ProductAddToCart` with the cart Game Object. Finally, pass the name of the product to your `ProductAddToCart` script, so that we can send this back to our `WebAPI` to indicate which product the user has purchased.

Similar to the `ProductRemover` class, create another script `ProductAddToCart.cs` inside the `Scripts` folder. The class must inherit from the `InputClickHandler` interface and implement the `OnInputClicked()` method as shown as follows:

```
public string addToCartProduct;
public void OnInputClicked(InputEventData eventData)
    {
        APIConnector.Instance.PurchaseProduct(addToCartProduct);
    }
```

Here in an one-liner code block, which just call the *Instance* of `APIConnector` that we already created and call the method `PurchaseProduct()` by passing the product name.

The following is the new snippet that you need to add in the `APIConnector`. The API Connector class just send a post `webRequest` to the services end points with product name.

```
public void PurchaseProduct(string productName)
    {
            UnityWebRequest webRequest =
              UnityWebRequest.Post("http://<<servicename>>
              .azurewebsites.net/API/Purchase/AddToCart",
              productName);
    }
```

 `UnityWebRequest.Post()` to send form data to a server via HTTP POST. In this case we are sending a Post call to our asset Web API with product name.

Changing pictures for Wall Paints

In case of **WallPaint**, we will have an additional toolbar that shows a **Refresh** icon. When users tap on that button, it will change the image inside the frame. This also shows that not only can you download and render a hologram, you can also update dynamic Holograms.

Find the **Refresh** Game Object from the toolbar control and then attach the `ChangePicture` script as we did for previous two cases. From the current script, pass the current object as reference to the `ChangePicture` class.

```
GameObject refreshObject =
          toolBar.transform.FindChild("Refresh").gameObject;
ChangePicture changePicture = refreshObject.AddComponent
          <ChangePicture>();
changePicture.changeImageObject = launchedObject;
```

Similar to the other two classes for toolbar control, create another script `ChangePicture` inside the `Scripts` folder:

```
public GameObject changeImageObject;
public void OnInputClicked(InputEventData eventData)
    {
            WWW www = new WWW("https://ch8assets.blob.core.
              windows.net/3dassets/painting_a.png");
            StartCoroutine(DownloadAssets(www));
    }

    private IEnumerator DownloadAssets(WWW www)
    {
```

```
yield return www;
if (www.error == null)
{
        Destroy(changeImageObject.
          GetComponent<Renderer>().material.mainTexture);
        changeImageObject.GetComponent<Renderer>().
          material.mainTexture = www.texture;
}
}
```

This code works in a similar way to how we downloaded the assets earlier. In this code, we refer on image asset directly from our storage, then use the similar approach to download the asset. Once the asset is downloaded, we remove the earlier texture object and apply the newly download texture.

 `WWW.texture` returns Texture2D generated from the downloaded data.

Well, now you have downloaded the product and have a toolbar attached to it; using that, you can **remove**, **refresh** or add them in to your *cart*. But one thing is missing here! This is simple but important enough to have it. Moving your object. You downloaded a TV, but you must need to place it on appropriate position in your wall or table. That's how you will get a feel of how it looks in reality. Let's have a look how we can achieve that.

Placing the holograms - spatial mapping

Add the following line of code with in the `DownloadAndProcessAssets()` method itself, just post the when `launchedObject` initialization. This add a `BoxCollider` component to the `launchedObject`, so that you can gazed on it. Post that, Add the `WorldAnchorManager` and `TapToPlaceScript` with the object. These two scripts are part of the `HoloToolkit` and you can find them in `HoloToolkit` | **Utilities** | **Scripts** and `HoloToolkit` | **SpatialMapping** | **Scripts** respectively.

```
BoxCollider boxCollider = launchedObject.AddComponent<BoxCollider>();
boxCollider.size = new Vector3(0.50f, 0.50f, 0.25f);
launchedObject.AddComponent<WorldAnchorManager>();
launchedObject.AddComponent<TapToPlace>();
```

In the Unity Editor, go back to the *Root* object in Object Hierarchy. Attach the `WorldAnchoreManager` with *Root* object as well.

 The `TapToPlace` class enable users to move objects and place them on real world surfaces. Attach this script on the object that you want to move. `TapToPlace` also use `WorldAnchor` component to enable persistence on the surface, which is supported by the `WorldAnchoreManager` class.

Finally, from the `Assets` folders, navigate to `Holotoolkit` | `SpatialMapping` | `Prefabs`, and drag and drop the `SpatialMapping` prefab to **Object Hierarchy**.

With that, when you tap on a menu item and a product is downloaded, by clicking on the toolbar you can take several actions. However, by taping on the actual product holograms, you can actually place them where you want to place it in actuality and visualise.

Finally, this is how the entire `DownloadAndProcessAssets()` method looks. This method does a major part of your application, starting from downloading the assets, rendering them as Holograms, attaching controls on the fly, and acting for each control.

```
public IEnumerator DownloadAndProcessAssets(WWW www, string assetName)
{
            yield return www;
            bundle = www.assetBundle;
    if (www.error == null && bundle != null)
    {
    launchedObject = Instantiate((GameObject)bundle.
        LoadAsset(assetName));
    launchedObject.transform.position = new Vector3(launchedObject.
    transform.position.x, this.gameObject.
    transform.position.y - .25f, this.gameObject.
    transform.position.z - 1.5f);

    BoxCollider boxCollider = launchedObject.
        AddComponent<BoxCollider>();
    BoxCollider.size = new Vector3(0.50f, 0.50f, 0.25f);
    launchedObject.AddComponent<WorldAnchorManager>();
    launchedObject.AddComponent<TapToPlace>();
    GameObject toolBar = Instantiate(Resources.Load("ProductTools",
      typeof(GameObject)) as GameObject, new
      Vector3(launchedObject.transform.position.x + 0.5f,
      launchedObject.transform.position.y,
      launchedObject.transform.position.z), Quaternion.identity);

    if (launchedObject.name != "painting_a(Clone)")
    {
```

```
        toolBar.transform.FindChild("Referesh").
            gameObject.SetActive(false);
    }
    else
    {
        GameObject refreshObject = toolBar.transform.FindChild("Refresh")
            .gameObject;
        ChangePicture changePicture = refreshObject.AddComponent
            <ChangePicture>();
        ChangePicture.changeImageObject = launchedObject;
    }

    GameObject removeProduct = toolBar.transform.FindChild
        ("Close").gameObject;
    ProductRemover remover = removeProduct.AddComponent
    <ProductRemover>();
    remover.toberemovedObject = launchedObject;

    GameObject cart = toolBar.transform.FindChild("AddToCart").
    gameObject;
    ProductAddToCart prr = cart.AddComponent<ProductAddToCart>();
    prr.addToCartProduct = this.launchedObject.name;
    }
    else
    {
        Debug.Log(www.error);
    }
}
```

Well, with that we are done with complete development process of our hologrpahic retail solution.

Deploying your app on a device

By now you are familiar with different processes of deployment of your holographic app in to your device.

Use any of the deployment approach and deploy your HoloRetail application into HoloLens and run the application. You will find a main menu appearing in front of you. Gaze into any of the product items; you should be able to visualize indicator and tap one of product to get them downloaded. Tap and place them anywhere in the room, use the toolbar to remove the product from the scene or add them to the cart.

Make sure your device is connected to the internet and able to access the Asset API Services. You can check the network from the Start menu, Select **Settings** | **Network & Internet.**

Extending the current solution

The objective of this project was to build an enterprise scenario with HoloLens with integrated cloud platform, and having both way communication with application and services. We also focused on how we can build a lightweight application by dynamically downloading assets and making them interactive on the fly. We have covered several interesting topics throughout this exercise; however, there are several enhancements that you can do to extend this solution, here are few of them.

Dynamic menu for products

This project has a fixed number of menu items (*Three - TV, Light and Wall Paint*); however, to extend your application and to match you real time need you can have make this menu dynamic with variable number and several types. You can achieve this by simply having another service call during the application launch which loads the list of products and generates the menu objects accordingly.

Adding Voice Command

You can extend this application by adding Voice Command. Extension scenarios could be-- say out a command to display a TV and your holographic app download and display the TV. You can perform tap and place to move it, however you can remove the object or add to cart using Voice Command. You may wonder how the voice command will work when there are multiple objects. In that case, you can find out which object is gazed when the Voice Command was raised, and based on that, you can take action.

Application and storage access security

At this point of time, the application is not secure. The Web API is exposed to the public and the Blob Storage Container has public access control. Anyone who hits the Service URL will get the JSON Data back, and if you access the Blob Asset URL, you can download it in the browser itself. There are several ways we can secure this application. We have used access token mechanism to secure both the Web API and access the Blob storage URL. We can define the Storage Access Policy for Blob Storage and user to only access URL when the container access level is granted.

Rotating the individual product

While you can do tap and place to move a product object from one place to other, you can enable rotating for an individual product to fit in a specific angle. You can have it rotate by using a separate rotate control, or add an additional button in the control toolbar to rotate in a specific angle.

Summary

This chapter covers an end-to-end integrated holographic application for a retail scenario. In this chapter, we covered how to create reusable Asset Bundles in Unity and take them to Azure Blob Storage. We set up the backend Azure infrastructure by creating a Azure Cosmos DB--SQL DocumentDB and referring the Blob Storage assets reference and then exposing the data using a Web API hosted as Azure App Service. The holographic apps consume those services and download the assets from Azure Blob Storage on demand and render them as holograms. We have also learned how we can attach toolbar for each product and take certain actions on the toolbar buttons. With all that covered, this chapter builds up a solid foundation for building different connected scenarios application with unlimited possibilities using HoloLens and Microsoft Azure stacks. With this knowledge, you can now develop an application that is lightweight, dynamic in nature and scalable in different aspects of several enterprise customer scenarios. With that, we have completed all the planned projects for this book. But, as we said earlier, the possibilities are unlimited with HoloLens. In the next chapter we will explore some more possibilities and scenarios with HoloLens to build a different level of Immersive experience app.

9
Possibilities

Over the course of the earlier chapters, we explored the various aspects of building holographic applications. We also developed several end-to-end solutions, starting from exploring HoloLens as holograms, a complete **Internet of Things** (**IoT**) scenario for the remote monitoring of buildings, and finally, building enterprise retail customer solutions. With that, we completed all of the three projects that we planned to develop in this book. While the developed application scenarios cover several technical integrations with HoloLens and cover different industry verticals, it doesn't mean that it's all that you can do with HoloLens. We have lots and lots of possibilities that you can achieve through HoloLens.

In this chapter, we will consider some of these possibilities and scenarios. The objective of this chapter is not to discuss step-by-step development code; rather, we will walk through the different possibilities along with their architectural and design aspects. In this chapter, we will discuss the following:

- HoloLens and Microsoft Bot Framework
- HoloLens in healthcare
- HoloLens and live streaming
- Handling secure keys with Azure Key Vault

HoloLens and Microsoft Bot Framework

A bot (meaning software robot) is a type of program that imitates human responses, basically used in scenarios for chat or offering support to users. Microsoft Bot Framework provides server-side functionality to easily implement these software bots for various purposes. In this section, we will see how you can leverage Microsoft Bot Framework and integrate it within holographic applications.

In the case of HoloLens, there can be multiple scenarios where a bot can be very useful, such as education or providing help to users during enterprise scenarios. Let's assume that you have developed a holographic application that will be leveraged by teachers and students to study new topics. Now, if any question comes up while learning this new topic, rather than looking for a physical person to ask this question, you can plan to leverage Bot Framework to implement a bot solution, which can offer virtual help to students.

Visualizing and interacting with bot along with primary Hologram

Solution

Microsoft Bot Framework provides different ways to connect with a bot and leverage its services. One of the way it exposes is through Direct Line API, which is a REST-based API for connecting with a bot directly. Direct Line API provides the following functionalities:

- Client authenticate, which is an authentication mechanism for calling applications
- Functionality to start conversations and send messages through HTTP POST
- Functionality to receive messages through HTTP GET

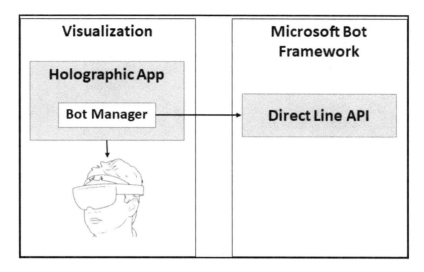

Design for calling Microsoft Bot Framework

Within your holographic application, you can write a sample class that can be responsible for all conversations with Direct Line APIs, to authenticate calls, start conversations, and get responses. The upcoming sections explain how to implement these calls.

Client authentication

Authentication is a two-step process. The first step is to get the conversation token using a secret key, and the rest of the conversation APIs are called using that token till the conversation is closed or the life of the token times out. For the next conversation to start, the application should get a new token. To get the token, all the application needs to do is call the tokens/conversations API using a secret key, such as the following:

```
-- connect to directline.botframework.com --
POST directline.botframework.com/api/tokens/conversations HTTP/1.1
Authorization: BotConnector[Secret Key]
```

Start conversation

Using the `token` received in the preceding step, you need to call the start conversation API to get a new conversation ID, as follows:

```
-- connect to directline.botframework.com --
POST directline.botframework.com/api/conversations HTTP/1.1
Authorization: Bearer [Token]

-- response from directline.botframework.com --
HTTP/1.1 200 OK
{
  "conversationId": [Conversation Id],
  "token": [Token],
  "expires_in": 1800
}
```

Send message

Using the `conversation Id` received in the preceding step, you can send multiple messages and receive responses:

```
-- connect to directline.botframework.com --
POST directline.botframework.com/api/conversations/[Conversation
Id]/messages HTTP/1.1
Authorization: Bearer [Token]
{
  "from": "Manish",
  "conversationId": "[Conversation Id]",
  "text": "sample text",
}

-- response from directline.botframework.com --
HTTP/1.1 204 No Content
```

Receive message

The next step is to receive conversations from the service. To do that, make a GET call to the conversation API, as shown:

```
-- connect to directline.botframework.com --
GET directline.botframework.com/api/conversations/[Conversation
Id]/messages HTTP/1.1
Authorization: Bearer [Token]

-- response from directline.botframework.com --
HTTP/1.1 200 OK
{
  "messages": [{
    "conversation": "[Conversation Id]",
    "id": "[Conversation Id]|0000",
    "text": "hello",
    "from": "Manish"
  }, {
    "conversation": "[Conversation Id]",
    "id": "[Conversation Id]|0001",
    "text": "Nice to see you, user1!",
    "from": "bot1"
  }],
  "watermark": "[watermark code]"
}
```

As you can see, it's quite a simple API call to build a conversation bot. In addition to these, you can use the TextToSpeech class of HolograpicToolkit to build in-voice conversational capability within the holographic bot.

 Refer to the Microsoft Bot Framework documentation at https://docs.bo tframework.com/ for more details.

HoloLens in healthcare

Healthcare is important for everyone, and if we bring the latest technology to the healthcare industry, that will be helpful for everyone. So, in this scenario, we will integrate HoloLens with HealthVault. HealthVault is a data platform for both personal, as well as business, enterprises. Users can use HealthVault to store and share their health information. Healthcare enterprises can use this centralized data store to build personalized services.

Scenarios can be: medical practitioner uses holographic application that connects and pulls data from HealthVault to visualize a patient's progress and health in one complete view, rather than going through multiple medical reports.

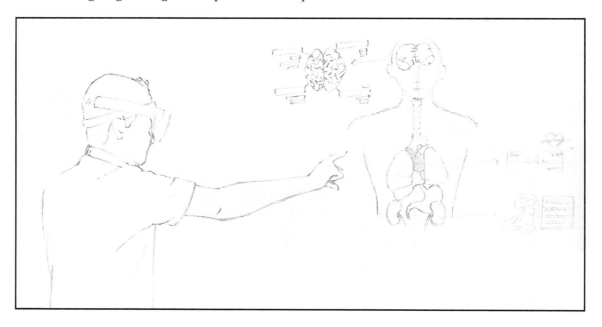

Visualizing patient health data using HoloLens

Solution

To achieve this scenario, solution design is broken down into two different layers. Within the first layer, we develop proxy services or APIs, which connect with HealthVault using HealthVault APIs, aggregate, and transform data as per holographic application consumption. The second layer is the holographic application itself, which consumes aggregated data from proxy services and displays results within the holographic application in a three-dimensional mode. This scenario can also be extended where two different medical practitioners can collaborate and discuss the same patient's health data fetched from HealthVault.

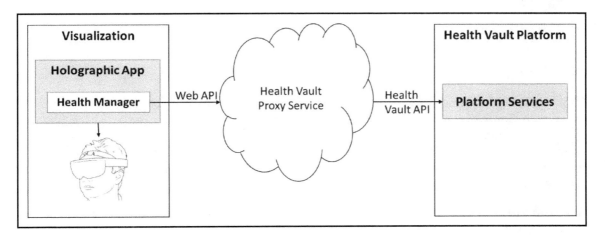

Holographic application consuming data from HealthVault

Search data from HealthVault

This is a sample code snippet to connect with HealthVault using its SDK and make a search query for a patient's weight history:

```
HealthRecordSearcher recordSearcher =
PersonInfo.SelectedRecord.CreateSearcher();
HealthRecordFilter weightFilter = new HealthRecordFilter(Weight.TypeId);
recordSearcher.Filters.Add(weightFilter);
HealthRecordItemCollection weightItems =
recordSearcher.GetMatchingItems()[0];

//weightItems is a collection of the patient's weight historical data.
   Further on, you can parse collection like following, to pull time
   based weight data for usage.
for each (Weight weightItem in weightItems)
```

```
{
    //weightItem.When - will give you the date
    //weightItem.Value - will give you the patient's weight at the
        specific date
}
```

Similarly, you can pull other health data using HealthValue APIs and share or reuse it within a holographic application.

 You can refer to HealthVault documentation at `https://www.healthvaul` `t.com` for more details.

HoloLens and live streaming

Sports and entertainment are two of the industries where HoloLens and MR can play a major role. Some of the mixed reality scenarios in the sports industry can be around players' training and game simulations. In the entertainment industry, the holographic scenario can be around content visualization. However, one scenario that is common in both industries is the live streaming of events and videos. There are different ways to consume videos and visualize videos, and HoloLens is one of them.

The scenario could be that a live sports or entertainment event is happening and it's being captured or recorded by 360-degree cameras. This live streaming of video can happen using Azure Media Services, and distant users can have a near real-time live experience of viewing that video on HoloLens.

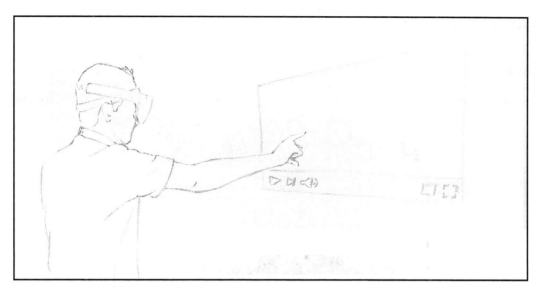

Visualizing live streaming using HoloLens

Solution

To achieve this scenario, we can leverage Azure Media Services to set up the streaming channel, which will be used to publish video output, encode video, and publish video streaming. After that, we will look into how this video streaming channel can be consumed within the holographic application.

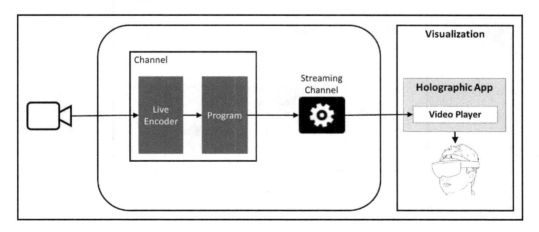

Visualizing live streaming of video on HoloLens

Setting up media services for live streaming

The first step to implement the solution is to set up a live streaming channel and, for that, follow the given steps:

1. Create a new **Azure Media Services** account within your **Azure Subscription**.

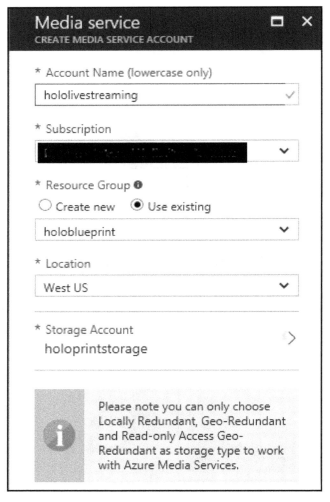

Create new Azure Media Service Account

2. After the new Media Service account is created, select the **Live Streaming** option and create a new channel for live streaming.

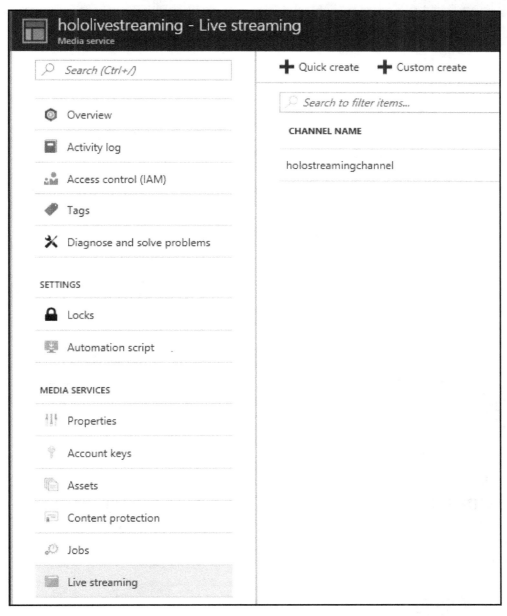

Create new live streaming channel

3. Select the newly created live streaming channel and start the **Live Event** within it.

Holostreamingchannel CHANNEL		Default LIVE EVENT	
⚙ Settings ⊘ Off Air → Live event ■ Stop ↻ Reset ⋯ More		■ Stop ⊘ Unpublish ▶ Watch 🗑 Delete	
Overview		**Overview**	
STATE	Running	NAME	default
INGEST PROTOCOL	RTMP	DESCRIPTION	
ENCODING TYPE	Pass Through	STATE	Running
PREVIEW URL	▓▓▓▓▓▓▓	ARCHIVE WINDOW	8 Hours
INGEST URL (PRIMARY)	rtmp:// ▓▓▓▓▓	ASSET NAME	holostreamingchannel-default-1491629298545
INGEST URL (SECONDARY)	rtmp:// ▓▓▓▓▓		
		Locators	
Live events		LOCATOR TYPE	URL
NAME STATUS ASSET ARCHIVE WINDOW PUBLISHED		ⓘ Streaming	http:// ▓▓▓▓▓
default ✓ Running ▓▓▓ 8 Hours ⊙			

Start the live streaming channel

4. From the preceding step, note down the following URLs:
   - **Ingest URL**: This is to be used by the video ingestion application
   - **Streaming URL**: This is to be consumed by the live viewer, that is, the holographic application in our case

There are many video streaming applications available; one of them is **Wirecast**, which you can download, configure the Streaming Channel Ingest URL within, and start live video ingestion to the media services channel. Use the following URL:

https://www.telestream.net/wirecast/overview.htm.

Consume live streaming within holographic application

Now we have video ingestion, and the streaming channel is in place. The next step is to consume this live streaming within the holographic application. To do that, you can follow this code snippet:

```
public class LiveStreaming : MonoBehaviour {
    public WWW webObject;
    public string videoUrl = "
       http://[mediaservicesaccount].streaming.mediaservices.
```

```
        windows.net/[....].ism/manifest ";
    public GUITexture guiTexture;

    //On Start, create WWW object and link it with Steaming URL
    void Start() {
        webObject = new WWW(videoUrl);
        guiTexture = GetComponent<GUITexture>();
        guiTexture.texture = webObject.movie;
    }

    //On Update, update movie texture and play
    void Update() {
        MovieTexture movieObject = guiTexture.texture as MovieTexture;
        if (!movieObject.isPlaying && movieObject.isReadyToPlay)
            movieObject.Play();
    }
}
```

Handling secure keys with Azure Key Vault

In the preceding scenarios, such as Bot Framework integration or HealthVault integration, we dealt with keys such as bot secret keys, which we don't recommend embedding or storing within a Holographic UWP application. For any enterprise scenario, these should be kept in secure locations, such as the Azure Key Vault solution for storing keys securely.

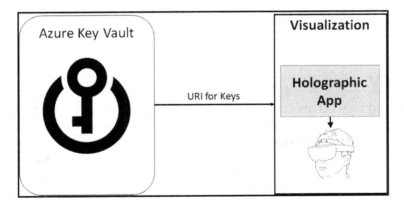

Key vault integration with Holographic application

In this section, we will look at ways to access Azure Key Vault from the holographic application. If you are not familiar with Azure Key Vault and would like to learn from the start, visit https://docs.microsoft.com/en-us/azure/key-vault/.

Azure Key Vault provides REST APIs to access secret keys and manage the overall Key Vault. For access keys, you must first get yourself authenticated and get an access token. All requests to Azure Key Vault must be authenticated through Azure Active Directory access tokens.

To retrieve the secret key from Azure Key Vault, you just need to make a GET REST API call with the secret name and secret version details:

- Replace *{secret-name}* with the name
- If you want to retrieve the current secret version, exclude the *{secret-version}*:

```
-- connect to Azure Key Vault --
GET /secrets/{secret-name}/{secret-version}?api-version=2016-10-01

-- response from directline.botframework.com -- HTTP/1.1 200 OK
{
  "value": "[secret value]",
  "id": [secret id],
  "contentType": [content type of the secret],
  "attributes":
{
      "recoveryLevel": [Purgeable | Recoverable+Purgeable |
         Recoverable, Recoverable+ProtectedSubscription],
      ......
},
  "tags": [],
  ......
}
```

Summary

In this chapter, you learned about the different possibilities in various industries with HoloLens. The first possibility was the scenario where you learned about leveraging Microsoft Bot Framework to develop self-learning applications for HoloLens. In the second scenario, you learned about connecting a holographic application with the HealthVault platform, which can change the way medical practitioners virtualize patient data. The third scenario was about connecting a holographic application with live video streaming channels.

The next chapter is the last chapter of this book, in which you will learn about device management for HoloLens devices in the enterprise space, along with other tips and tricks that will help you develop holographic applications.

10
Microsoft HoloLens in Enterprise

In the course of this book, you have learned about HoloLens as a device, developed enterprise-ready holographic applications, deployed them on a device, and learned to connect holographic applications with the cloud platform. In the previous chapter, you learned about the various possibilities with holographic applications and HoloLens as a device, but the possibilities are unlimited. In this chapter, we'll learn more about enterprise readiness toward HoloLens device management and holographic application management. So, in this chapter, we will discuss the following three topics:

- Microsoft HoloLens Commercial Suite
- HoloLens device management using Microsoft Intune
- Installing, publishing, and distributing applications across HoloLens devices

Microsoft HoloLens Commercial Suite

Microsoft delivers HoloLens devices in two models: Development Edition and Commercial Suite. Commercial Suite is recommended for enterprises that require enhanced features for device management capabilities and security. The following are some of the additional features included within Microsoft HoloLens Commercial Suite:

- **Mobile Device Management (MDM) for HoloLens using Microsoft Intune**: An MDM solution like Microsoft Intune can be leveraged to manage HoloLens devices, such as managing device settings, managing security settings, and installing applications remotely.

- **Windows Store for Business**: Enterprise IT department used to have an enterprise private Windows Store, where they managed their company-specific applications. The same store can be leveraged for managing enterprise-specific holographic applications also.
- **Data Security**: The commercial edition of HoloLens supports data encryption using BitLocker, so now you can secure HoloLens devices just as you would secure any other Windows device.
- **Identity**: The commercial edition supports device credentials using Azure Active Directory. So, enterprises already using Azure Active Directory can simply start using these devices using the Azure Active directory credentials.
- **Work Access**: The commercial edition can now be connected through enterprise WiFi networks, which are secured through credentials or certificates. Also, commercial version devices now support **Virtual Private Network** (**VPN**) connectivity, so users can connect to Enterprise VPN networks using a HoloLens device.
- **Kiosk mode**: Similar to the Windows tablet kiosk mode feature, you can now have kiosk mode on the commercial version of the HoloLens device, where you can restrict the number of applications that can run when the device is in kiosk mode. This feature is especially helpful when giving demos to external audiences.

Feature	Commercial Edition	Development Edition
Mobile Device Management (MDM)	✓	✓
Windows Business Store	✓	
Enterprise WiFi access (based on certificate)	✓	
Identity – Login with Azure Active Directory	✓	✓
Identity – Login with Microsoft Account	✓	✓
Data Security – Device Encryption (BitLocker)	✓	
Virtual Private Network (VPN)	✓	

Feature comparison between commercial and development editions

Microsoft Intune is a cloud-based MDM service that enterprises which provide management services of PCs (Personal Computers), mobile devices, and mobile applications use. The primary features provided by Microsoft Intune are as listed:

- Manage mobile devices and PCs that are used for accessing company data
- Manage mobile applications used by company employees
- Enforce company security requirements on devices and applications and make them compliant as per the enterprise security policies

For details, refer to `https://www.microsoft.com/en-us/cloud-platform/microsoft-intune`.

HoloLens device management using Microsoft Intune

Microsoft Intune is one of the MDM solutions available that can be leveraged to manage Microsoft HoloLens for managing general and security settings and rolling out applications on multiple devices. If an enterprise already has Microsoft Intune set up, there are two ways to register a device:

- Manually register the device on Intune
- Automatically register the device on Intune

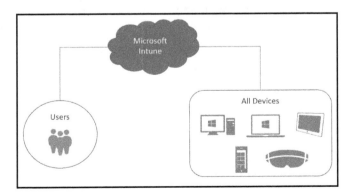

Microsoft Intune to manage enterprise devices and applications

Manual registration

When a HoloLens device is not registered on Intune, perform the following steps to register:

1. Start the device.
2. Select the **Settings** application.
3. Within the **Settings** application, select the **Accounts** section.
4. Within the **Accounts** section, select **Work access**.
5. Now, select the **Enroll into device management** option:
   * This will prompt you to enter your organizational account. Sign in using your organizational credentials.
6. If authentication using your organizational account is successful, enrollment to Intune is successfully completed.
7. Restart the device so that all the organization certificates and policies get applied to the device.

Automatic registration

For automatic registration, your organization should be using Intune and Azure Active Directory, and have automatic enrollment of Windows 10 devices set up within Intune using the Azure Active Directory credentials. If that is already set up, then all it requires is that you log in to your HoloLens device using your organizational credentials, and this will automatically trigger the registration of the device on Intune.

 Refer to the following article to enable Windows 10 to automatically enroll onto Intune using Azure Active Directory-- `https://docs.microsoft.co m/en-us/intune/deploy-use/set-up-windows-device-management-wit h-microsoft-intune#azure-active-directory-enrollment.`

Installing applications on HoloLens

After you have set up your devices for the enterprise, the next step is to install and distribute holographic applications onto those devices. As these target devices are enterprise devices, there are now multiple options to install or push applications to these devices. The simplest way is to use Windows Device Portal, as you have seen earlier in this book. The other way to install an application is to use Microsoft Store for Business.

Device Portal for application deployment

To deploy an application using Device Portal, the first thing you should have is an application package and all its dependencies packages. If you have packages, perform the following steps:

1. Start Device Portal.
2. Navigate to the **Apps** tab within Device Portal.
3. Within the **Install App** section, click on the **Browse** button under the **App package** subsection.
4. Select application package from the network path.
5. Also add dependency packages if there are any.
6. Click on **Go**; this will install the application package.

Business Store for application deployment

Business Store is a dedicated, isolated, and private store for an enterprise. Enterprises can store various types of applications on this business store, and a holographic application is one of them.

Now, if you need to download any application from the Business Store, start the Store application within HoloLens. There, you will find a section for your organization; select that and you will find a list of applications hosted specific to your organization. Select application, download, and install.

 You can visit the Microsoft Windows holographic application public store at `https://www.microsoft.com/en-us/store/collections/hlgettings tarted/hololens`.

Summary

Through this book, you have learned to develop your first holographic application, develop an enterprise-ready connected application, deploy it on actual devices, and finally, some of the possibilities with holographic applications. During the course of this book, you developed three holographic applications. You started with the first project, in which you developed a standalone holographic application, created 3D Assets, applied gesture scripts, and learned to deploy it on a device. This was followed by a second project where you integrated IoT solutions with holographic applications, and learned to connect holographic applications with backend Azure services. In the third and final project, you developed an end-to-end retail solution as a holographic application, followed by an evaluation of some of the possible scenarios in which holographic applications can play major roles, and also, how an enterprise can manage HoloLens as a device. This is not the end; this is just the beginning of your journey toward MR.

Index